An Introduction to Architectural Conservation

Philosophy, Legislation & Practice

Nick Lee Evans

RIBA ✛ Publishing

Published by RIBA Publishing
part of RIBA Enterprises Ltd, The Old Post Office, St Nicholas Street, Newcastle upon Tyne, NE1 1RH

ISBN 978 1 85946 529 5

Stock code 81930

British Library Cataloguing in Publications Data
A catalogue record for this book is available from the British Library.

Publisher: RIBA Publishing
Commissioning Editor: Sharla Plant
Production: Michèle Woodger
Cover design: Michèle Woodger
Book design: Academic + Technical, Bristol
Typesetting: Academic + Technical, Bristol
Printed and bound by CPI Group (UK) Ltd, Croydon, CR0 4YY

RIBA Publishing is part of RIBA Enterprises Ltd.
www.ribaenterprises.com

CONTENTS

ACKNOWLEDGEMENTS

I would like to thank Bob Franklin, Peter Bell, Karen Banks and David Stewart for their advice and encouragement, and my wife Kay for her patience and editing.

FOREWORD

by Robert Franklin

"Conservation is about the building, not the architect".
Alex Bulford, RIBA Conservation Registrar.

There are approximately half a million Listed Buildings in Britain, over a million buildings in Conservation Areas and, within the national building stock, more than five million houses predating 1919. In addition, some 45% of the entire national building industry turnover is spent on existing buildings, giving cash value of some £6 billion. There are therefore many reasons for architects to seek appropriate information and training in materials and techniques that are usually very different from those used in more modern constructions. This guide's aim is to help the practitioner towards a fuller understanding, and to make necessary and appropriate design and technical judgements, by providing an overview of current conservation principles and how they are applied to older buildings.

This introductory guide is therefore partly aimed at the generalist who occasionally works on older buildings, and who has not yet decided to become an accredited specialist, or sees no reason to become so. These architects may well be technically skilful and have a high level of knowledge in their local vernacular and traditional building methods, but may seek more detailed guidance on how to approach the conservation of a building under statutory protection of Listing, especially in light of recent changes to planning law and guidance.

This becomes very much more important when we are dealing with, for example, the survey, repair and maintenance of a cathedral; or we update and extend a listed house, re-order a church or convert a historic warehouse to a new use. We are therefore concerned with best practice when working on 'historic buildings', rather than the merely 'old', and this is the over-riding concern in assessing applications to the RIBA Conservation Register in order to ensure that Listed buildings are in safe hands.

So the primary purpose of this guide is, as the title suggests, is to assist those architects embarking on the more demanding work to buildings that have an established cultural value of Architectural or Historic Interest, and therefore have legal protection against damage or inappropriate change.

It is inescapable that, for architects, the difficult area of 'design' with its burden of indefinable creative, subjective and emotional imperatives must also be addressed. One definition of 'conservation' indeed, is 'the management of change', and change means design. In the Listed Building context therefore, it must first be accepted that ego is best left at the door, and the spirit and cultural value of the place – and to a high degree, engagement with the mind-set and skills of the original architects and builders – are fully absorbed. Only then can any success in the outcome be judged, and to what degree reticence, circumspection and self-effacement has been allowed to illuminate the process.

Thus the critical difference between new-build and conservation is that context and the constraints offered by existing buildings require a different mind-set to that applied to new buildings. But given that, it is creativity and the 'architectural mind' that is applied to work on historic buildings just as much as it is to entirely new structures. Knowledge of all periods of design style, three-dimensional vision and understanding of structural performance, construction methods, materials science and building behaviour are all as critical to the invisible repair and stabilisation of a medieval tower or cathedral vault, say, as to the addition of new work within that tower or vault, or to the creation of an entirely new building adjacent to it.

Conservation and adaptation of great architecture from the past is therefore not about market share, or expanding a business, or 'making a mark', and it is about much more than simply following approved processes for repair and preservation, for it includes providing creative, subtle and reversible ways to overcome defects in the original construction which have led to accelerated decay, and in making the changes that are essential to maintain a building's useful purpose for today, and thus its reason to continue existing. Style is therefore less critical than design quality within the chosen style, and much less important than fitness for purpose, sound execution, craftsmanship and beauty.

Historic buildings still have much to teach us about sustainability too, and much current research is suggesting that their thermal performance and natural ventilating properties are significantly better than previously thought. It could indeed be said that there is nothing more sustainable, and educational, than an historic building used as simply as possible. Their numbers are also very small in relation to the entire national building stock, so whilst it is prudent to improve them so far as is technically appropriate without causing harm to their historic significance, the known risks of damaging rather than improving performance suggests a light-touch, or changing the way we occupy old buildings, rather than relying on the 'techno-fix' and enthusiastic modification in ignorance of possible damaging consequences.

This guide is therefore an introduction that will help architects new to this field, to navigate the cultural, legislative, aesthetic, technical and, to an extent, even the moral issues that will be encountered in this important field. Becoming the 'safe pair of hands' that historic buildings deserve and that Grant Aid providers properly require, is therefore a goal worthy of pursuit in its own right.

Robert Franklin RIBA SCA, Former Chair RIBA Conservation Group and
RIBA Council Member

FOREWORD
by Peter Bell

"These old buildings do not belong to us only; … they have belonged to our forefathers and they will belong to our descendants unless we play them false. They are not … our property, to do as we like with. We are only trustees for those that come after us."
Words spoken by William Morris in 1889 which resonate in the development of conservation philosophy throughout the twentieth century and the conservation movement of today.

Whether we recognise it or not, architectural heritage plays an important role in our national identity, our cultural consciousness and in the quality of our lives. We value buildings for different reasons, some of which are very personal: sentiment, pride, nostalgia, antiquarianism, patriotism or the desire for familiar reference points in a changing world are just some. The significance and relative value we attach to historic buildings changes with time and from individual to individual depending on upbringing, education and personal preferences, but take away the medieval parish church, the cherished street scene or the old Victorian boot scraper and to some greater or lesser extent our lives would be diminished.

Few would challenge the premise that we should conserve the significant buildings of previous generations, or that future generations will conserve the best of our generation's buildings. However, the way in which we preserve buildings can prove controversial.

Unsurprisingly, there is no shortage of opinions about which buildings we should protect and how we should go about preserving, conserving or restoring them. We have a sophisticated plethora of statutory protection for almost every type of heritage asset: we have processes, procedures and policies all aimed at protection and the positive management of change. We have no end of restoration programmes on television and resources online. Conservation principles however are still poorly understood by many, and there is often misunderstanding or distrust between those charged with conserving our heritage and those who own it, appreciate it or actively campaign for it.

Tony Aldous once quoted Lewis Caroll's *Through the Looking Glass* when he said that conservation is one of those humpty dumpty words 'when I use a word' said Humpty Dumpty, '*it means just what I choose it to mean - neither more nor less*'.[1] It is the arbitrary 'humpty dumpty' approach to architectural conservation that gives the conservation professions and practitioners a bad name and causes some people to take the opinion that conservation philosophy is at best subjective or at worst a dark art. The only way to move away from that position is through careful study and understanding, education and better communication. The architectural student as well as the seasoned practitioner

1 Caroll, L. (2007), *Through the Looking Glass*, London: Penguin Classics

needs to acquire a thorough understanding of what makes buildings significant, and understand the effects that our interventions have on them today and in the future. A confused understanding of conservation philosophy, legislation and practice leads to poor conservation of buildings and in the worst case to irreversible harm or total loss.

We have a remarkable architectural heritage which grows richer and more diverse with every passing decade. Some buildings are at risk through neglect or lack of funding, but the greatest risk to our heritage is the ill-conceived repair or alteration carried out without a proper understanding or consideration. The use of cement mortars and impervious membranes in historic buildings is not a phenomenon which passed with the twentieth century. Well-intentioned but often ill-considered repairs, alterations and extensions are regrettably still common place. We need to continually learn from our past mistakes and rise to the many challenges that caring for buildings in the twenty-first century brings. I have the greatest respect for Nick Lee-Evans' conservation work, having worked alongside him on many projects over 25 years. Nick was one of the first architects to become an RIBA Specialist Conservation Architect and I believe that this book will be an indispensable tool for the seasoned professional as well as the student.

This introduction to architectural conservation should be welcomed as it addresses complex issues in a concise and readable way. It comes some 130 years after William Morris founded the Society for the Protection of Ancient Buildings and a century after monuments were first afforded statutory protection by the Ancient Monuments Consolidation and Amendment Act of 1913. This is the first time in a generation that such a comprehensive guide to conservation philosophy, legislation and practice has been published, and as a conservation practitioner of 30 years, I highly recommend it.

Dereliction and ill-considered repairs and 'restoration' may not be as widespread as when William Morris founded the Society for the Protection of Ancient Buildings, but the same threats are ever present. I hope that all the trustees of our historic buildings will benefit from the pages which follow. This book should give those with responsibility for architectural conservation, owners, architects and conservation practitioners, a deeper understanding of the subject and a passion to care for and conserve our architectural legacy.

Peter Bell
Conservation Officer

CHAPTER 1
THE ROLE OF THE ARCHITECT IN MAINTAINING AND ADAPTING HISTORIC BUILDINGS

Architects and surveyors play a vital role in both repairing and maintaining historic structures, as well as in making careful changes and **conversions** for modern occupation. Renewed usefulness through occupation is not the only value these buildings hold, they also provide a physical record of our history and often hold an emotional meaning in a community derived from its collective experience of a place. They give important insights into our past that can be physically experienced and understood, more so than through records on drawings, film or in words. They remind us of how earlier generations lived, and provide our historic towns and villages with the architectural heritage that is part of our cultural identity. They also often symbolise wider values, such as faith, and by definition listed buildings possess an architectural and historic importance. Historic structures for these reasons are valued as single examples of fine architecture or as groups that create urban landscapes of high **cultural significance**. Historic buildings are a treasured non-renewable resource, and therefore should only be altered after due consideration of their **significance**.

Heritage assets

Listed buildings, scheduled ancient monuments, historic gardens and castles, **conservation areas** and historic landscapes are given a special status as **heritage assets** in law because they hold meaning for society beyond their functional utility. Because of this cultural significance, strong feelings are often aroused in the community when we seek to alter such buildings. Government therefore seeks to protect them through legislation, and in consequence requires an informed approach in order to obtain consent to change designated heritage assets.

To work successfully on listed buildings and to ensure **interventions** are not damaging, architects need to understand the legislation and the **conservation** principles that underpin them. Such knowledge will allow architects to understand the objectives of conservation officers and other guardians of our heritage, and therefore facilitate in tailoring applications to satisfy their requirements. By developing these skills, architects will be better placed to advise clients with greater authority, make successful listed building applications, use time more efficiently in the process and leave behind better preserved built heritage for future generations.

In this guide we examine the principles behind the legislation as well as the designation systems currently in force to protect our nation's heritage assets.

Preserving character does not mean preventing change

Buildings are not static objects; they decay, and their use often changes over time. Weather will transform them, and without **maintenance** and **renewal** buildings fall into disrepair. This can happen surprisingly quickly; tiles slip, gutters block and damp enters into the building leading to rot and fungal decay.

Architects and surveyors who are often involved in the maintenance of buildings need to recognise the causes of decay and know how to rectify problems. The choices made however when making **repairs** can also affect the architectural, historical and cultural significance of the building. For example, renewing all the decayed stone tracery in a church window will lose its **patina** and sense of age, whereas a more careful and thoughtful approach to specification will aim to preserve as much original material as possible, resulting in better retention of the essential character of the window and its architectural and historic interest. The critical judgement is in balancing these values against the need to avoid further work in the near future.

Old buildings can be of great architectural, cultural or historical significance, and in order for them to be maintained there has to be a sustainable continuing use which will pay for and ensure their upkeep. Often such buildings become redundant when they no longer suit their original uses, needing sensitive **alteration** to ensure their survival. The design skills of conservation architects can ensure a building has a continued use without destroying its architectural, historical or cultural value. This guide recognises the need for change, and sets out to give an introduction to the philosophical and legislative framework which should guide those who seek to work on historic and listed buildings.

Building technologies and sustainability

There isn't scope in this guide to explain how all of the various and numerous historic building materials behave. Such knowledge is paramount for architects to bring to conservation projects, and there are many sources both printed and online that can guide architects, including SPAB Technical Guides, and publications from English Heritage on the repair of traditional building materials and techniques. This book however seeks to explain the commonly accepted principles of conservation and show how they can be applied to good working practices on a project, from the initial analysis of the **fabric**, through to the choice of repair, specification writing and contract management, and will also direct the reader to sources of specialist advice, in the Further Reading section.

The technical challenges of maintaining and adapting historic buildings require special knowledge by architects and surveyors, and furthermore an understanding that historic buildings were created with technologies that were usually based on local natural resources; often, such materials perform in very different ways to our modern construction methods. Materials and construction techniques used in traditional buildings are characterised by the widespread use of materials derived from natural sources such as lime, clay, stone, oak etc. These materials are normally more permeable than modern building materials and thus allow moisture to migrate through the building fabric more easily.

It is only from around 1919 onwards that new materials such as reinforced concrete, Portland cement mortars, damp proof membranes and related techniques such as cavity wall construction came into common use. Modern construction techniques also rely on impervious barriers such as plastic membranes to make the building resistant to moisture, often coupled with mechanical ventilation and heat recovery. Modern architects are not normally trained to understand traditional building and construction technologies, and if modern construction methods and materials are applied to an older building, damage can occur for example through moisture entrapment and reduced ventilation[1].

All historic buildings have also taken considerable resources to build, and their replacement would require high levels of energy and resources in materials manufacture, transport and construction. There is therefore a strong case to retain and use or re-use our older building stock simply on grounds of sustainability, although there is a growing need to reduce the energy used to light, heat and maintain them. Upgrading old and significant structures to make them more energy efficient involves a special, and growing skillset. The standard methods of upgrading buildings, such as reducing air admittance, double glazing windows and using vapour impermeable insulation can cause many further problems and lead to accelerated fabric decay in historic buildings.

Do you need to be a specialist?

When work is proposed on the most important historic buildings there is clearly an advantage in being an accredited conservation architect who can demonstrate specialist knowledge. English Heritage promotes accredited conservation architects, and looks to registration bodies such as the RIBA and AABC to identify and accredit those with the appropriate skills. English Heritage and the Heritage Lottery Fund (HLF) ensures that agents applying for their grants are accredited, and the Government is promoting that all work on publicly owned historic buildings should use accredited specialist architects, surveyors and engineers. The steady growth of conservation accreditation for professionals does not stop many non-specialist, non-accredited architects, doing valuable work on listed buildings, most of which are private domestic buildings, as long as due care is taken and recognition of being out of one's depth leads to seeking specialist advice and researching the problem area when necessary.

Increasingly those architects who have built up their skills to work on historic buildings may find it an advantage to have their skills recognised through accreditation bodies such as the RIBA Conservation Register, especially as more clients are now seeking architects with such additional qualification, and more public sector clients, heritage trusts and church authorities adopt membership on the Conservation Register as a prerequisite of tendering.

The RIBA Accreditation Scheme is based on recognising different levels of skills, so architects can develop over time from Registrants to Conservation Architects (CA) to Specialist Conservation Architects (SCA). It is hoped many architects will come to recognise the advantages of doing so. Those thinking of applying to enter the RIBA's Conservation Register can find more details on the RIBA website[2].

In the recession following the banking crisis of 2008 the number of new buildings architects have been asked to design has significantly dropped. However, RIBA benchmarking surveys show that since 2008, retrofitting and reuse of older buildings has been a growing proportion of architects' workload[3]. There is clearly an ongoing need for those architects who can creatively maintain and adapt historic structures to continue their use and be made more energy efficient. This guide should be of interest to architects and surveyors seeking to exploit the work available not only in conserving and adapting historic buildings, but also those involved with refurbishing and retrofitting.

Summary

This book seeks to explain the commonly accepted principles of conservation and show how they can be used to guide good working practices on a project from the analysis of the fabric, choice of repair, dealing with legislation, to specification writing and contract management, and will also direct architects to sources of specialist advice.

It is written for non-specialist architects as well as conservation professionals. It seeks to introduce the commonly accepted principles of conservation, and explain how these have led to our current legislative framework concerning historic buildings, allowing the reader to understand the aims and objectives of conservation officers, statutory amenity societies and other guardians of our heritage. An understanding of the commonly accepted conservation principles allows architects to advise clients accurately, make successful listed building applications, and ensure that interventions do not unnecessarily damage historic fabric.

Finally, this guide aims to provide a practical outline to good working practice; taking the reader through the process of a listed building application, preparing the tender package and administering the appropriate form of contract. It will highlight the differences in approach when working on historic building projects compared to normal projects.

Notes

1   Thomas, A., Williams, G. & Ashurst, N. (1992), *The Control of Damp in Old Buildings*, SPAB.
2   www.architecture.com
3   Colander, (2011/2012, 2013/ 2014) *RIBA Buisness Benchmarking*, RIBA.

CHAPTER 2
THE UNDERLYING PRINCIPLES OF CONSERVATION

Historic buildings and building fabric can as easily be lost through ill-considered alteration and over-zealous restoration, as through demolition. In almost all cases, it is a loss which cannot be reversed. Over-restoration can easily destroy the character and interest of a building by replacing so much material that it becomes a parody of the original, whilst insensitive alterations and additions can destroy the scale and design **integrity** of the original building.

When working on historic buildings it is often a challenge to know when to stop, or how much to replace. This raises questions as to whether it is better to repair using alien or novel materials, how much should be done with new or salvaged materials, where salvaged materials have been sourced from, and whether the repair be invisible or visible. These are ethical questions, because when working on fragile historic buildings we have a responsibility as guardians of our heritage to pass these structures on to future generations in a form that allows them to continue to be read and re-examined in their context of time and space, with their original design integrity intact. Architects and other construction professionals working on historic buildings need to consider these questions. Indeed practitioners who embark on conservation work without considering them are likely to experience difficulties in representing their clients and obtaining consents, and may even risk damaging the fabric they seek to preserve.

A conservation officer or English Heritage inspector will use the current principles of good conservation practice as a framework by which to judge applications. When intending to work on historic buildings it is therefore imperative to develop an understanding of the principles, policies and guidance available in order to work within this framework. Once the language and principles are understood, the architect will be able to argue his/her case and limit the frustrations that can arise when those granting consents don't agree with the proposals. "Why won't they let me make a new doorway in this wall?", or "Why can't I remove this structure?" will then become questions you need not ask, or alternatively such changes can be justified by the architect's research and reasoning.

Historically, many of the principles of good conservation practice have been expressed and set out in various ways (please see 'Further Reading'). The most important may be considered to be:

*   The SPAB Manifesto of 1877
*   The first Ancient Monuments Act of 1882.
*   International Charters and Guidelines particularly those published by the International Council on Monuments and Sites (ICOMOS)
*   English Heritage guidelines
*   Current legislation in the Planning (Listed Buildings and Conservation Areas) Act 1990, the National Planning Policy Framework 2012.

The SPAB manifesto

Debates about the correct ways of preserving heritage have exercised minds for millennia. For our purposes, we can start with an 1840s RIBA statement on how ancient buildings should be protected, which is still hard to find fault with today. This RIBA statement was powerfully reinforced in the second-half of the nineteenth century when design philosophers and practitioners such as William Morris helped halt the fashion for **restoration** that sought to re-create a perfect gothic architecture through conjectural replacements that in turn destroyed fine work of the 'wrong' period. Georgian or early Renaissance architecture was not so widely valued, and many architects sought an alternative aesthetic perfection rooted in an idealised view of a medieval Christian past. Morris and his fellow campaigners were appalled to see historic architectural features removed from churches and cathedrals to make them more 'perfect'. For example in 1832 at Canterbury Cathedral, for stylistic reasons, a medieval Romanesque tower at the west end of the church was demolished and replaced with a replica of a gothic tower on the other side of the west end, in order to make the building symmetrical.

In 1877 Morris initiated the Society for the Protection of Ancient Buildings (SPAB)[1] and published a manifesto which gave a compelling expression of what have become the basic tenets of conservation work:

- We are guardians of the ancient buildings we inherit, for future generations. We should not feel free to do with them as we please now.
- Maintenance by effective and honest repair, which is clearly distinguishable from the original, should be the first consideration.
- We should never do more work than is necessary to preserve the building. The original manifesto also espoused the view that it was better to build a new building than alter or enlarge an old one to suit modern use. However SPAB now has dropped this principle and accepts that it can be necessary to adapt historic buildings to ensure they have a use, which will help ensure their continued maintenance. The Society has further refined their guidance and they now define their purpose on their website as below[2]:

- **Repair not restore**
  Although no building can withstand decay, neglect and depredation entirely, neither can aesthetic judgement nor archaeological proof justify the reproduction of worn or missing parts. Only as a practical expedient on a small scale can a case for restoration be argued.
- **Responsible methods**
  A repair done today should not preclude treatment tomorrow, nor should it result in further loss of fabric.
- **Complement not parody**
  New work should express modern needs in a modern language. These are the only terms in which new can relate to old in a way which is positive and responsive at the same time. If an addition proves essential, it should not be made to out-do or out-last the original.
- **Regular maintenance**
  This is the most practical and economic form of preservation.
- **Information**
  To repair old buildings well, they must be understood. Appreciation of a building's particular architectural qualities and a study of its construction, use and social

development are all enlightening. These factors also help us to see why decay sets in and how it may be put right.

- **Essential work**
  The only work which is unquestionably necessary (whether it be repair, renewal or addition) is that essential to a building's survival.

- **Integrity**
  As good buildings age, the bond with their sites strengthens. A beautiful, interesting or simply ancient building still belongs where it stands however corrupted that place may have become. Use and adaptation of buildings leave their marks and these, in time, we also see as aspects of the building's integrity. This is why the Society will not condone the moving or gutting of buildings or their reduction to mere façades. Repairs carried out in place, rather than on elements dismantled and moved to the work-bench, help retain these qualities of veracity and continuity.

- **Fit new to old**
  When repairs are made, new material should always be fitted to the old and not the old adapted to accept the new. In this way more ancient fabric will survive.

- **Workmanship**
  Why try to hide good repairs? Careful, considered workmanship does justice to fine buildings, leaving the most durable and useful record of what has been done. On the other hand, work concealed deliberately or artificially aged, even with the best intentions, is bound to mislead.

- **Materials**
  The use of architectural features from elsewhere confuses the understanding and appreciation of a building, even making the untouched parts seem spurious. Trade in salvaged building materials encourages the destruction of old buildings, whereas demand for the same materials new helps keep them in production. The use of different but compatible materials can be an honest alternative.

- **Respect for age**
  Bulging, bowing, sagging and leaning are signs of age which deserve respect. Good repair will not officiously iron them out, smarten them or hide the imperfections. Age can confer a beauty of its own. These are qualities to care for, not blemishes to be eradicated.

International charters

Twentieth century conservators have considered in great depth the consequences of their actions, and debated how best to work on a building or object through international meetings and subsequent charters setting down codes of best practice. These principles are now used to guide our actions, whether training professionals and conservators or preparing a specification for restoring an old building.

The pre-eminent of these charters and codes of practice belongs to the International Council on Monuments and Sites (ICOMOS), now the widely agreed and orthodox code used by those involved with historic building conservation. ICOMOS brings together approximately 10,000 professional members involved in many aspects of conserving heritage from all around the world. ICOMOS is recognised by UNESCO and advises on the selection of World Heritage Sites in addition to establishing best practice for conservation in its many forms.

It is the major global, non-government organisation of its kind, and is dedicated to promoting the application of theory, methodology, and scientific techniques to the conservation of architectural and archaeological heritage. The principles it promotes were enshrined in 1964 in a charter adopted at a conference in Venice on the Conservation and Restoration of Monuments and Sites (the Venice Charter). These codes of practice were further developed following a long battle in Australia over the future of the old outback mining town of Burra. It was one of few intact surviving mining towns and had a series of structures and buildings from the pioneering mining days of the nineteenth century, but the town had been built on land sacred to Aboriginals. The authorities wrestled with the dilemma of whether to preserve the mining town or return the landscape to the earlier sacred condition of the Aboriginals. The debate revolved around issues of relative significance, when it was appropriate to restore and by how much. In the end the government decided to preserve the mining town rather than recreate the Aboriginal landscape. The principles on which this decision was made, and the language and definitions used were set down in the Burra Charter produced by Australia's national committee of ICOMOS[3].

The ICOMOS Burra Charter is one of the clearest codes of practice used in modern conservation. Understanding its principles allows architects working on historic buildings to develop an understanding of sound conservation behind sound conservation practice. It forms the basis for informing sound decisions about how much to repair, alter, or replace.

Current British planning legislation is designed to protect our historic environment and **natural heritage**. It acts the internationally agreed set of principles that have been developed by bodies such as ICOMOS since 1964. UK governments have signed up to various international charters (Granada 1985[4], Valletta 1992[5]) which require them to protect our built heritage and implement these principles in laws and policies.

On the ICOMOS website architects can find copies of various charters relating to good conservation practice including the Athens Charter of 1931, the Venice Charter of 1964, and the Burra Charter of 1981, revised in 1999.

The European Convention for the Protection of the Architectural Heritage of Europe (Granada 1985), and The European Convention for the Protection of the Archaeological Heritage (revised) (Valetta 1992) can also be found at its website[6].

The ICOMOS guidelines

The ICOMOS criteria are expressed succinctly and in a way most relevant to building professionals in 'Guidelines for Education and Training in the Conservation of Monuments, Ensembles and Sites' (1993). This code is currently used by the RIBA Conservation Register and others, as the test for competence which needs to be demonstrated through evidence in a portfolio of work and case studies when applying to become an accredited conservation architect. This competence demands the ability to:

[a] Read a monument, ensemble or site and identify its emotional, cultural and use significance;
[b] Understand the history and technology of monuments, ensembles or sites in order to define their identity, plan for their conservation, and interpret the results of this research;

[c] Understand the **setting** of a monument, ensemble or site, their contents and surroundings, in relation to other buildings, gardens or landscapes;

[d] Find and absorb all available sources of information relevant to the monument, ensemble or site being studied;

[e] Understand and analyze the behaviour of monuments, ensembles and sites as complex systems;

[f] Diagnose intrinsic and extrinsic causes of decay as a basis for appropriate action;

[g] Inspect and make reports intelligible to non-specialist readers of monuments, ensembles or sites, illustrated by graphic means such as sketches and photographs;

[h] Know, understand and apply UNESCO conventions and recommendations, and ICOMOS and other recognised Charters, regulations and guidelines;

[i] Make balanced judgements based on shared ethical principles, and accept responsibility for the long-term welfare of cultural heritage;

[j] Recognise when advice must be sought and define the areas of need of study by different specialists, e.g. wall paintings, sculpture and objects of artistic and historical value, and/or studies of materials and systems;

[k] Give expert advice on maintenance strategies, management policies and the policy framework for environmental protection and preservation of monuments and their contents, and sites;

[l] Document works executed and make same accessible;

[m] Work in multi-disciplinary groups using sound methods;

[n] Work with inhabitants, administrators and planners to resolve conflicts and to develop conservation strategies appropriate to local needs, abilities and resources.

Even if an architect does not intend to become an accredited conservation architect the key to good practice when working on historic buildings is to apply sound conservation principles of the kind outlined by ICOMOS. If they are used in the preparation of documentation, then the architect provides him/herself with a sound basis for discussion with the officials regarding the merits of their schemes. The ICOMOS guidance is examined in more detail below:

[a] Read a monument, ensemble or site, and identify its emotional, cultural and practical significance

The ICOMOS Burra Charter defines the significance of an object as the 'aesthetic, historic, scientific social or spiritual value for past present or future generations'[7]. Therefore since changing historic buildings will inevitably impact on their architectural, cultural or historic significance, it is essential to investigate and understand that significance before any changes are proposed. It is also important to understand what is meant in the Charter by the word 'conservation' – the Australian, ICOMOS definition is 'all processes of looking after a place so as to retain its cultural significance'[8].

The charter highlights as its first principle the need to be able to read a historic building in order to identify its significance. When an architect 'reads' a building they study it at many different levels:

• to identify its age and its style, and examine how it has been altered and extended over time;

• to consider how it has been constructed and examine how it has been used;

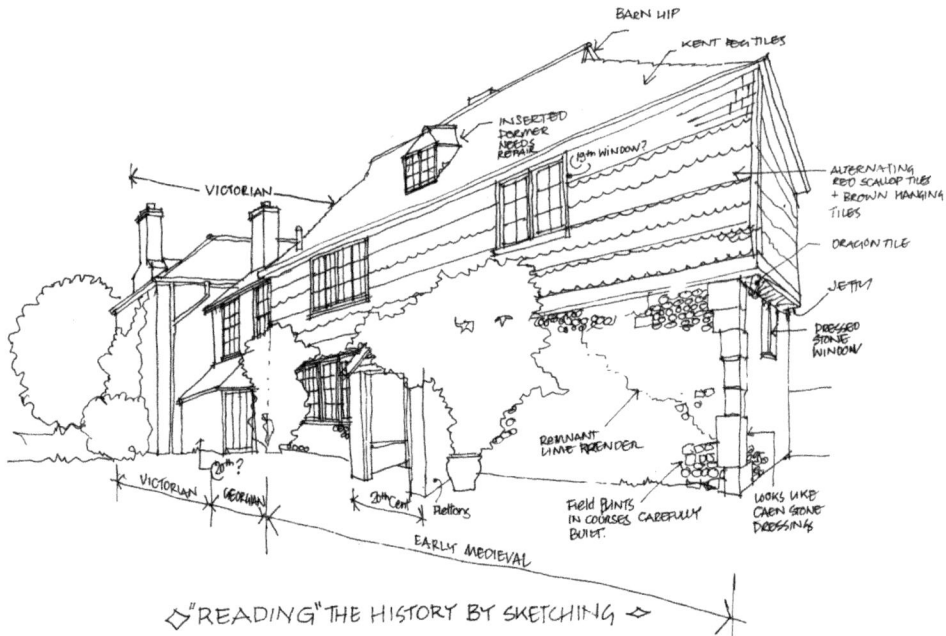

FIGURE 2.1 A hand drawn sketch made on site where notes of the first 'reading' were made for latter reference in the office.

- analysis is carried-out on the materials used and how they interact;
- to discover as much as they can about its history and who used it; and to look at its relationship to the site and surrounding buildings.

Architects who have not taken the time to read and understand the historic building risk not only damaging significant historic fabric, but also wasting time designing inappropriate alterations. It is unprofessional to propose a scheme that is unlikely to get consent, or for others to recognise significance in any part of the fabric that the architect hasn't yet grasped.

For most historic buildings, 'reading' times can vary depending on size, complexity, age, degree of change since first constructed, and the 'reader's' experience. Many conservation architects find the act of sketching a building helps them to gain more intense insight into the building's history. See Figure 2.1. Practice and experience of reading many historic buildings eventually enables a preliminary reading to be taken almost at a glance.

[b] Understand the history and technology of monuments, ensembles or sites in order to define their identity, plan for their conservation, and interpret the results of this research

Books are readily available on the history of most forms of vernacular buildings in Britain, and advice and guidance can be obtained from the major statutory amenity societies such as the Victorian Society, SPAB and the Georgian Group – see Further Reading for details. Many specialist building forms such as water and windmills and older industrial buildings have dedicated preservation societies full of knowledgeable experts who can guide an architect through basic research. Local conservation officers will also have an understanding of the local technologies, whilst traditional building forms can be very helpful in identifying rare and unusual features, as well as helping to put the building into the context of wider regional variations.

[c] Understand the setting of a monument, ensemble or site, their contents and surroundings, in relation to other buildings, gardens or landscapes

Historic buildings can also have value as part of a group. Some may not be very significant in their own right, but with their neighbours they can create an attractive and culturally significant space. In an old village the group of medieval cottages grouped around the ancient church is often a locally valued space. It can be read as an expression of the history of the community, valued emotionally for the events that took place there in the past, as well as being valued aesthetically as an attractive space rich with the patina of ages. These prized building groups often contain simple cottages, which if altered significantly could destroy the harmony and character of the landscape.

It is important to understand a building's place in the spectrum of similar buildings by age and by use. To do this, comparison to other similar buildings may require the architect to carry out some general research into the history and construction technology of similar structures and typology. This can help place the architect's understanding of the building into a wider context and help to identify whether it has rare or unique features that need carefully planned conservation. Conversely, it can help demonstrate that the historic building is of lesser significance against other similar examples.

[d] Find and absorb all available sources of information relevant to the monument, ensemble or site being studied

The architect will need to search all possible archives to find and study any available documentation, including any recent site history held in the local authority planning department, and make a close visual examination of the building and its curtilage. It is advised to also consult the local conservation officer and, in Grade II* and above buildings, the regional English Heritage Building Inspector for insight and an early indication of the likely response to proposals. Talking with owners and neighbours can also give an understanding of the building's local emotional significance.

This examination and on-site 'reading' should yield a deep understanding of the building, which will inform the approach to conservation and change. Likewise, depending on size and complexity, site and building measured surveys should not be delegated to others unless it is to obtain specialised digital data. In this case, once the document is received the conservation architect should use it as a template on which to overlay the result of a close personal inspection and analysis of every part of the building.

It is important when working on historic buildings to seek out sources and guides not only on the general building typology, but also sources related to individual buildings, to inform an evaluation of the building's architectural, historical, cultural and aesthetic significance. These sources provide information that the physical examination and 'reading' of the building will not fully yield. When starting work on a historic building it is good practice to start by assembling as many sources as possible. Interested owners often have collections of information such as old pictures and plans, and local guide books.

For example local libraries and diocesan records can yield valuable information. Very often the owners of a property are very willing to assist in the hunt to find out information sources. This method of discovery can also be helpful as a means of educating owners into the significance of their property. Each listed building has a description of its significance held by English Heritage or their equivalent national authority, as well as the local authority Planning office (now easily accessed on the web). These are helpful as a starting point, but descriptions can be very short and often only highlight the items deemed significant during an often cursory external inspection conducted at the time of listing.

Other archive sources to consult are discussed further in Chapter 6, but they will include Borough, County Council and Unitary Authority **Historic Environment Records** (HER), Victoria County History (British History On-Line), National Archives, Ordinance Survey historic maps, county and district council archives (photos in particular), national and local museums, the internet, The Pevsner Architectural Guides to the Buildings of the British Isles and national and local libraries for books containing reference to specific buildings, styles, materials, building techniques, and social history.

Without a clear understanding that has come from an architect's research into the building's history, it is impossible to begin to weigh up the significance of various parts of the structure and put together a clear argument to defend the design.

Conservation officers will normally be very cautious about allowing significant changes to historic buildings and can err on the side of preservation rather than alteration. Their understanding of the history and significance of the building will often come from the architect's research, which should therefore be as comprehensive as possible. As a basic principle, once the history study and statement of significance is completed, the conservation architect should know more about the building than anyone else.

[e] Understand and analyse the behaviour of monuments, ensembles and sites as complex systems

All buildings are 'complex systems' created from many components that are interdependent. Architects design modern buildings with elements that are designed and known to work together, but historic fabric can often behave in unexpected ways. As construction professionals, architects are trained to understand the use of modern building materials and manufacturers' research, who publish the behaviour of their products in different circumstances. Old buildings are often constructed from locally produced building materials, of inconsistent quality, which can behave erratically. There can also be complex interactions of historic materials which can lead to forms of decay that are sometimes hard to predict. The way many historic materials generally decay and interact is now better understood following years of academic research. It is therefore critical, when working on a historic building, for the architect to familiarise themselves with any available research into the properties of traditional materials prior to specifying repairs. SPAB and English Heritage have produced many useful guidance notes on most forms of traditional materials, and these are listed under Further Reading.

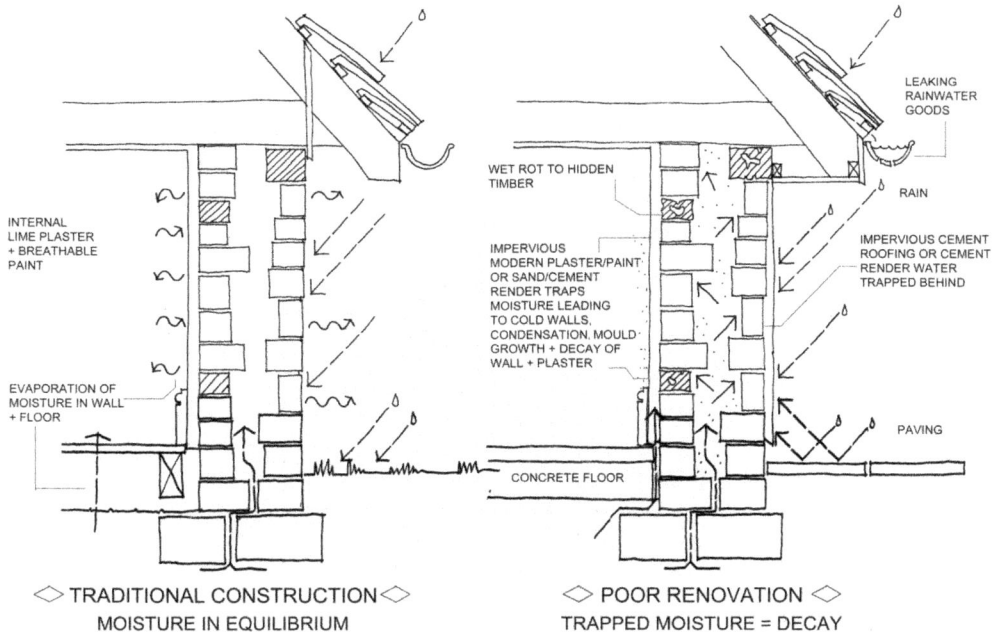

INTERNAL
LIME PLASTER
+ BREATHABLE
PAINT

EVAPORATION OF
MOISTURE IN WALL
+ FLOOR

WET ROT TO HIDDEN
TIMBER

LEAKING
RAINWATER
GOODS

RAIN

IMPERVIOUS
MODERN PLASTER/PAINT
OR SAND/CEMENT
RENDER TRAPS
MOISTURE LEADING
TO COLD WALLS,
CONDENSATION, MOULD
GROWTH + DECAY OF
WALL + PLASTER

IMPERVIOUS CEMENT
ROOFING OR CEMENT
RENDER WATER
TRAPPED BEHIND

PAVING

CONCRETE FLOOR

◇ TRADITIONAL CONSTRUCTION ◇
MOISTURE IN EQUILIBRIUM

◇ POOR RENOVATION ◇
TRAPPED MOISTURE = DECAY

FIGURE 2.2 Traditional construction and the problems associated with poor renovation.

The use of modern materials and building technologies in the repair of historic buildings often cause problems because the complex interactions of modern and historic materials have not been fully understood. For example, the inexperienced will often specify cement in mortars used to repair historic lime mortar masonry, as they are stronger and set quicker than the original lime mortars. However, failure to understand that the cement in the mortar prevents moisture vapour migrating through the joints in solid masonry, as it did with the original lime mortar, may in turn lead to a more destructive decay of the masonry due to entrapment behind the harder face of the stone or brick. The risk then is that a hard frost will freeze the trapped moisture and separate the face of the masonry from the body of the brick or stone.

There is still much to learn of the ways traditional materials perform, including in thermal performance, and the fact that historic structures have often survived so long is an indication that the materials and technologies that made them are robust, and should be respected.

Groups of historic buildings can also be complex systems creating local micro-climates that can funnel wind, shed water on their neighbours or shelter them. These factors should be taken into account before the architect considers changes and demolitions which can cause acceleration of decay by new variations in the local environment.

[f] Diagnose intrinsic and extrinsic causes of decay as a basis for appropriate action

Decay in all building materials is an inevitable process. Most materials, will have a limited life span and will decay at varying rates. For example, medieval iron work was made of 'puddled' iron, with most of the impurities worked or hammered out of it in smelting or on the anvil.

Such ironwork can last centuries unprotected, as can still be seen on numerous church doors. Much Victorian ironwork and steel, if left unprotected, can decay fairly rapidly as the iron oxidises where it is exposed to air.

Insects or moulds consume wet timber, and smoke sulphates and acid rain dissolve all kinds of stone. Some Victorian architecture was decorated by the mixture of contrasting stones, and limestone above sandstone often led to the sandstone surface dissolving. Architects working on historic buildings need to develop forensic techniques to establish the causes of decay and building failures: with a better understanding of the agencies of decay it is easier to rectify them successfully. The reader is advised to consult the numerous specialist texts[9] on materials performance in order to be able to draw sound conclusions on what is found in the building.

[g] Inspect and make reports intelligible to non-specialist readers of monuments, ensembles or sites, illustrated by graphic means such as sketches and photographs

Part of a professional's duty when working on historic buildings is to educate the owners on how to care for them in the future. Clearly explaining in understandable language the proposed changes and why they are being done can educate owners about the correct ways of maintaining the buildings in their care.

It is also the duty of a professional to record before and during alterations in ways that future owners, historians, and architects can easily interpret. These records can be an important information source to guide future owners and building professionals, and help them make decisions in the interests of preserving the historic fabric. They can also help future historians architects and planners interpret the building correctly, and show where repairs have succeeded or failed.

[h] Know, understand and apply UNESCO conventions and recommendations, and ICOMOS and other recognised charters, regulations and guidelines

The understanding of conservation principles as expressed in the various charters of ICOMOS and UNESCO[10] gives architects the philosophical understanding to make decisions about the historic buildings in their care, which can be defended to those charged with statutory protection of our heritage. A link to web sites for the relevant ICOMOS and UNESCO Charters and Conventions can be found on the website of COTAC at www.cotac.org.uk

[i] Make balanced judgements based on shared ethical principles, and accept responsibility for the long-term welfare of cultural heritage

It is possible to take the view that historic buildings are too precious to alter and therefore it is unethical to change them at all. This may of course be true of scheduled ancient monuments, but if most others are to survive they need to be used. From use comes value and if buildings are valued they will be repaired and maintained for the benefit of future generations. One of the roles of construction professionals is to balance the needs of clients, who will ensure the fabric is cared for, against the loss of irreplaceable heritage that they might otherwise like to make in order to maximise their financial benefit. It is an architect's ethical duty to establish as much as they can about the significance of the fabric before changes are proposed, and unless there is a good reason why the change is needed, the architect should leave the historic fabric unaltered. Architects should ensure that their decisions are based on a shared responsibility not only for their clients, but for future generations of owners and the wider community.

[j] Recognise when advice must be sought and define the areas of need of study by different specialists, e.g. wall paintings, sculpture and objects of artistic and historical value, and/or studies of materials and systems

Architects always have a professional duty to recognise the limits of their knowledge and skills, and to seek expert advice when necessary. Historic buildings are often complex systems and their maintenance and repair often needs specialist expertise beyond that of the architect. Repairs done by those without sufficient expertise can easily destroy historically significant fabric and therefore, part of the role of the conservation architect is to seek the right expertise. There are many knowledgeable specialist conservators working in very specific areas of building conservation, including wall paintings, textiles, sculpture and monuments, fibrous plaster conservation, historic paint and mortar analysis etc. Cathedral Communications annually publish The Building Conservation Directory (www.buildingconservation.com) which is a good source of specialist conservation experts, professionals and contractors.

[k] Give expert advice on maintenance strategies, management policies and the policy framework for environmental protection and preservation of monuments and their contents and sites

Even after restoration, old buildings will still need maintenance, and if inappropriately carried out it can easily destroy the significance of the building. Therefore architects and surveyors have a duty to educate clients with expert advice on the maintenance of their historic structures when opportunities arise. Advice can be given at the handover after completion of a project, or it can be given on a regular basis. Advice on maintenance strategies is a significant source of work for many conservation architects. For example most church denominations have adopted a system requiring that their buildings are inspected by a suitably qualified architect or surveyor every 5 years (known as quinquennial inspections) to assess the condition of the building and recommend what work is needed and how to prioritise repairs.

[l] Document works executed and make same accessible

The documentation of executed conservation works is good practice as it can help future conservators understand what is original and what has been repaired. Architects working on Anglican churches often find invaluable source material in the diocesan archives where drawings and specifications of alterations which were approved are normally stored.

These records are normally extensive for twentieth century works but may include records from earlier centuries. The information collected can inform decisions and help the assessment of the age and significance of the existing fabric. Researching listed building and planning consents in local authority archives can also provide useful information. Leaving a clear set of records of the work executed for future generations is therefore good practice. Many non-domestic buildings will now have ongoing health and safety files produced to comply with the legal requirements of the Construction Design and Management Regulations and updated each time there is construction work. These files may be a suitable place to leave records of any conservation or repair projects which have been carried out for the benefit of future owners.

[m] Work in multi-disciplinary groups using sound methods

It should not need saying that architectural professionals working on irreplaceable historic fabric need to use sound and robust methods and will need to co-operate effectively, taking specialist advice from experts, fellow professionals and craftsmen. Good practice will require listening and taking action on the advice given, as well as the development of a consensus on the best way forward.

[n] Work with inhabitants, administrators and planners to resolve conflicts and to develop conservation strategies appropriate to local needs, abilities and resources

Architects are trained to evaluate the needs of a client and develop built solutions that meet them. However, the importance of preserving historical fabric often requires clients and their architects to accept a less than optimal solution in order to preserve built fabric of significance. Skilled conservation architects use good design and communication skills to resolve conflicts between client's needs and the views of the administrators and planners whose legal role is to ensure any permitted changes sustain the significant fabric.

English Heritage guidelines

In 1983 the Historic Buildings and Monuments Commission for England (English Heritage) was formed to advise government and others, to help ensure the nation's heritage is protected for future generations. English Heritage is an important body staffed by experts in various conservation disciplines such as architects, planners, historians and archaeologists. It oversees the process for listing buildings and provides assistance and advice to local planning authorities as well as commenting on listed building applications on Grade I & II* buildings, and advises the Secretary of State for Culture Media and Sport before any work can be done to scheduled ancient monuments. It is currently funded partly by the government, and its publications are normally treated in the planning system as **supplementary planning guidance**, which local planning authorities routinely formally adopt as part of their planning policies. English Heritage publications[11] are often a major source of advice for building professionals and always reflect sound conservation principles derived from sources such as ICOMOS and the wider conservation sector.

After an extensive debate internally and externally in 2008, English Heritage published 'Conservation Principles Policies and Guidance for the Sustainable Management of the Historic Environment', as an expression of common principles held by the organisation and to strengthen the credibility and consistency of their decisions and advice. It is a useful reference document for anybody working in the field of building heritage conservation. Not only does it provide a useful set of definitions for commonly used terms but it also outlines six principles as a framework for the sustainable management of the historic environment:

1. The historic environment is a shared resource
2. Everyone should be able to participate in sustaining the historic environment
3. Understanding the significance is vital
4. Significant places should be managed to sustain their values
5. Decisions about change must be reasonable, transparent and consistent
6. Documenting learning from decisions is essential

English Heritage believes that the historic environment is a non-renewable economic, social and cultural resource which needs management. They accept that the historic environment is constantly changing and keeping significant places in use may require continual adaptation and change. However, they argue that every reasonable effort should be made to minimise the adverse impacts on significant places. They are not there to stop change, however they believe it can be necessary to balance public benefit of proposed change against the **harm** to the place. The assessment of a historic asset's significance is a vital part of their role in advising where change is being proposed. They also see that the weight given to heritage values needs to be proportionate to their significance. English Heritage also see part of their role as educating the public about the value of our heritage, as learning plays a central role in sustaining it for future generations.

Notes

1   Morris, W. (1877), *Manifesto for the Society for the Protection of Ancient Buildings*.
2   www.spab.org.uk
3   www.icomos.org/en/
4   Council of Europe (Granada 1985), *Convention for the Protection of the Architectural Heritage of Europe*, available from www.conventions.coe.int
5   Council of Europe (Valetta 1992), *Convention for the Protection of the Archaeological Heritage (revised)*, available from www.conventions.coe.int
6   www.conventions.coe.int
7   *Australia ICOMOS (2013), The Burra Charter: The ICOMOS Charter of Places of Cultural Significance*: Article 1.2, p. 2.
8   *Australia ICOMOS (2013), The Burra Charter: The ICOMOS Charter of Places of Cultural Significance*: Article 1.4, p. 2.
9   A good starting point for any research on materials performance can be found in books produced by English Heritage, and published by Ashgate Publications in the Practical Building Conservation series. Currently the series includes books on stone, timber, roofing, metals, glass and glazing mortars, renders and plaster.

10 UNESCO, (Paris 1972), Convention concerning the protection of the world cultural and natural heritage, available at www.whc.unesco.org
11 English Heritage publications are available at, www.english-heritage.org.uk/publications

Further reading

> *Australia ICOMOS (2013) The Burra Charter: The Australia ICOMOS Charter of Places of Cultural Significance*, also available from www.international.icomos.org and their UK website www.blog.icomos-uk.org.
> British Standards Institution (2013), *BS 7913 Guide to the Conservation of Historic Buildings*, London, BSI.
> Drury, P. (2011), Conservation: An Evolving Concept, *Building Conservation Directory 2011*, Tidsbury, Cathedral Communications Ltd, also available at www.buildingconservation.com.
> Earl, J. (2003), *Building Conservation Philosophy*, Shaftsbury, Donhead.
> English Heritage (2008), *Conservation Principles Policies and Guidance for the Sustainable Management of the Historic Environment*, London, English Heritage, also available from www.english-heritage.org.uk/publications.
> English Heritage (2013), *Conservation Basics*, Farnham, Ashgate Publishing Ltd.
> Forsyth, M. (2007), *Understanding Historic Building Conservation*, Oxford, Blackwell Publishing.
> Orbasli, A. (2008), *Architectural Conservation: Principles and Practice*, Oxford, Blackwell Science Publishing.
> Morris, W. (1877), *Manifesto, June 1877*, London, SPAB archives, also available in part from www.spab.org.uk
> Pendebury, J.R. (2009), *Conservation in the age of Consensus*, Abingdon, Routledge.
> Pevsner, N. et al. (1966 onwards) *The Pevesner Guides to the Buildings of England Series*, Yale University Press.
> HM Government (1882), *Ancient Monuments Act*, available from www.heritagelaw.org/AMA-1882
> COTAC, Understanding Conservation website available from www.understandingconservation.org. This is a very useful educational resource with, free to use, training units intended to challenge and develop the skills of those involved with building conservation. The website is based on the ICOMOS Education and Training Guidelines.
> COTAC, website available at www.cotac.org.uk. The Conference on Training in Architectural Conservation (COTAC) brings together a large number of organisations with the purpose of improving the standards of education for everyone involved in building conservation. Its web site has a very good digital bibliography with web links to many of the international charters on heritage protection which have been signed by the British Government.

CHAPTER 3
REASONS WHY HISTORIC BUILDINGS MAY NEED TO CHANGE

One of the key concepts when working on historic buildings, reflected in government policy and planning legislation, as well as accepted conservation practice, is that there should be no changes to the historic fabric unless it can be shown that there is a reason why change is needed, whilst care is taken to preserve the building's significance.

No buildings remain totally unchanged over time. Change comes to buildings in many ways: components decay over time, stone decays with acid rain attack, bricks and mortar erode, slates and tiles crack and delaminate, paint flakes off. Aside from weather and decay, buildings may become redundant and need new uses, for example vernacular farm buildings not suited to modern farm machinery are left empty until a new use can be found, old factories no longer suited to modern production techniques are abandoned. Even where the basic use made of a building remains constant over time, change and adaptation are needed due to the evolution of the way we live and work. There is a constant requirement to adapt our structures to suit our current lifestyles. Schools, hospitals and factories of the nineteenth century are not used unaltered. Virtually all Victorian buildings have had new systems for heating, plumbing and electricity inserted.

Heritage authorities believe their role is to manage these changes in order to minimise their impact and the loss of historical fabric. Conservation Officers will therefore be assessing proposals for change and balancing the need for change against the significance of the structure.

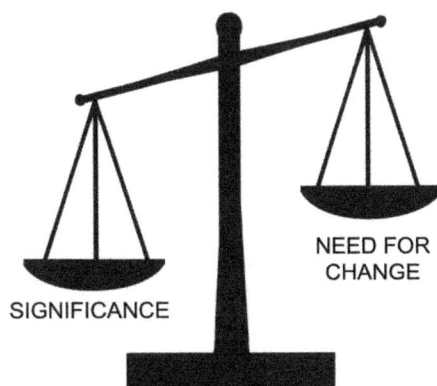

NEED FOR CHANGE

SIGNIFICANCE

FIGURE 3.1 It could be said that our heritage legislation seeks to balance the need for change against the value and preservation of historic, artistic and cultural significance of the fabric.

The main reasons why historic fabric can change are well established in British planning law and they include the need for:

[a] Investigation
[b] Repair
[c] Restoration
[d] Conservation
[e] Adaption and alteration
[f] To make a building more sustainable

The underlying conservation principles are now well established and when working on historic buildings and seeking listed building consent it is important to understand what a conservation officer means by these different terms and to understand when consent of alteration is needed:

[a] Investigation

Prior to undertaking a major set of repairs to a building, Consent may be justified as a piece of investigation into why a given problem is occuring. Often on historic buildings architects have to carry out this research before they can put together a full application for further changes. Normally there is an exchange of correspondence with the conservation officer who will agree to the investigative work being done without having to make a listed building consent application. The architect may need to supply a location plan for the works and some photographs and a short explanation why they are needed. However, occasionally the investigative works will be large enough to warrant the need for their own listed building consent application.

A common example of this type of investigation is where an architect suspects that a piece of walling is more modern than the surrounding fabric. By making small holes it is possible to establish that the walls have square sawn timber studs which are probably twentieth century in origin, or that the plaster is of a modern type rather than made of traditional materials dating from earlier centuries. If the walls are shown to be modern they are less significant and therefore consent is likely to be given to change them.

Another example of when investigation may be needed prior to the main listed building consent application is when building fabric failure occurs, the cause of which is not obvious, for example where cracks indicate there is some structural movement and its cause is not obvious because it is hidden by linings. It is therefore necessary to remove the outer linings to see inside the structure and establish the cause before specifying rectification measures.

Most projects on historic buildings involve some element of research and discovery as the project progresses, and it is very common to agree the repairs in principle on the listed building consent application and vary them with agreement with the conservation officer as the work of investigation proceeds. This can avoid the need for two applications, one for researching problems and a second for agreeing the work required.

The conservation officer will need to accept the investigation is justified and may need to know how any holes will be repaired. The size of any investigation will need to be minimised and targeted.

[b] Repair

Change may be needed to repair fabric which has decayed or failed. If the architect is repairing a listed building, listed building consent would not normally be required when the repairs are essential and done with the same materials and methodology, without changing the character of the building. In practice most repairs do involve some alterations to the character of the building, even if the same materials are used as the original, as the new material will not have weathered over time. For example a newly lime-pointed listed brick wall will look very different with its stark white mortar compared to the greyer weathered mortars.

Repairs are not normally contentious listed building applications, but they do need to be considered to ensure their impact is as minimal as possible. SPAB and the Listed Property Owners Club provide advice to owners of historic buildings on the correct ways to carry out fabric repairs, and SPAB also operate a phone and email advice service[1].

There can be occasions where more of the original fabric can be maintained if modern repair techniques are used such as resin bonding or steel reinforcement, and therefore a change in material is justified. A common example of this is where an old timber beam has rotted at its ends where it is supported on a wall, and the repair could be achieved by replacing the beam. However, the original beam can be kept by introducing a steel plate or an iron shoe bolted to the old timber, or introducing resin anchored with stainless steel rods to replace the rotted timber.

When repairing a building the emphasis is therefore on keeping as much original fabric as possible. Other common examples of repairs include:

- Piecing in new timbers where timber has decayed on a window, door, or rafter
- The use of steel plates to join timbers
- Repairing significant area of external renders etc.
- Cutting out stone or brick or a stone mullion to the glass line only where it has decayed, and pinning in a new stone piece to match the original
- On a stone structure the historic mouldings were normally shaped to shed water away from the building. Over time they decay and their repair can be justified to protect the rest of the building
- Replacing significant numbers of cracked or slipped roof tiles rather than the whole roof
- Repairing significant areas of wall finishes, using lime renders and traditional paints which breathe and let the wall dry out naturally (using modern paints often will be seen as an alteration)
- The permitted use of lead paints on Grade I listed buildings to match existing paints

It is also important to realise that historic repairs and alterations may be a significant part of the history of the building, and like the original fabric, they may be significant and need preservation.

Often necessary routine repairs do not require formal consent, unless they are to a scheduled monument. Consent however is needed if they change the character of the building, and this can be a matter of some debate. Therefore the architect would normally notify the conservation officer with details of his/her intentions, methods and materials, to establish whether a listed building consent will be needed.

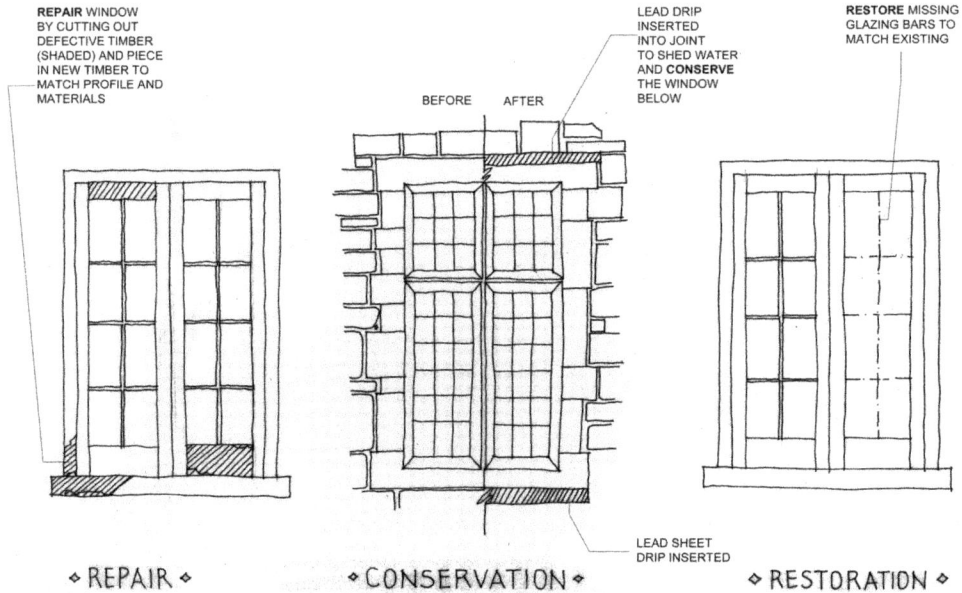

FIGURE 3.2 Three examples of permitted work to listed windows, the left hand side of the left window is being 'repaired'; the right hand side of the same window is having its glazing bars 'restored'; whilst the central window is having lead weathering strips added to 'conserve' the stone work below.

The terms '**repair**' and '**restoration**' can often be confused, and it is important when talking to conservation officers that the architect understands the differences. A repair is needed to preserve the fabric of the building, without which the building will decay. Restoration is not needed for preservation, it is done to enhance the building and replace an element which had disappeared.

[c] Restoration

Changes proposed to restore a missing element of the building can be amongst the most contentious applications. It is normally done for aesthetic reasons. Buildings evolve over time and the restoration of historic fabric tends to imply that the work of one period in a building's history is more significant than that of another. This is a **value judgement** open to debate.

Early nineteenth century architects 'restored' many churches back to their gothic medieval architecture, removing evidence of later alterations. It was seen as perfectly acceptable to remove 'modern' Georgian or baroque elements and add scholastically correct new elements in the gothic style. This approach is now considered unacceptable and certainly would not conform to the ICOMOS guidelines, since it involves the loss of considerable amounts of historic fabric. James Wyatt (1746–1813), one of the leading architects of his generation, set about 'improving' several cathedrals including Durham, Hereford, Lichfield, and Salisbury by removing elements from previous centuries and restoring gothic elements he felt had been lost. In France, Eugene-Emanuel Viollet-le-duc (1814–1879) 'restored' many medieval French monuments to an idealised form of gothic, including whole citadels such as at Carcassonne.

George Gilbert Scott (1811–1878) prolifically restored numerous churches across Britain. These restorations were often indistinguishable from the original medieval fabric and may have used recycled historic material. The Society for the Protection of Ancient Buildings (SPAB) was formed to oppose this approach to maintenance of historic buildings, promoting instead an emphasis on necessary repair and the preservation of historic fabric, and that where restoration was needed it should be done honestly and clearly.

This has led to an understandable resistance to conjectural 'restoration' in the minds of most heritage professionals since they believe it is not honest. There still however may be occasions when restoration can be justified complying with the best practice principles found in the ICOMOS charter. Examples of acceptable forms of restoration could include:

- A room may have a fine ornate cornice but one part is missing and its restoration can be justified to re-establish its continuity
- Restoring a cast iron gutter on a building by replacing an inappropriate plastic gutter which would have had cast iron originally
- Replacing a modern hard cement plaster and restoring with a lime based plaster, because it is not allowing the structure behind it to breathe, leading to damp and decay
- Restoration of a soft lime mortar to replace hard modern cement pointing which is leading to decay of the surrounding stone or brick
- Use of green oak timbers to repair missing rafters in an old oak barn to re-establish its continuity

It is important to remember when considering restoration that it is normally seen as good practice to subtly distinguish the original fabric from the restored fabric. This allows future generations to trace the history of the building accurately. The subtle insertion of new fabric is an art the architect working on historic buildings needs to cultivate. It can be one of the marks of a good conservation project.

[d] Conservation

Sometimes it is necessary to alter the fabric in order to conserve the fabric by preventing or slowing down decay. Accredited conservation architects will often use conservation techniques to slow **natural change**, decay and weathering of historic fabric. Examples of conservation based interventions include:

- The introduction of a piece of lead over a piece of medieval carving which is to be used to divert any rain water coming from above away from the delicate medieval stone, and slow decay caused by weathering
- The painting of a stone with a lime-based 'shelter coat' to slow decay and **consolidate** the surface
- Targeted cleaning to remove contaminants which are accelerating decay

[e] Additions and alterations

Each year English Heritage publishes a list called 'Heritage at Risk Register'[2], and in many of these cases the buildings are at risk because they are no longer used and therefore they are not maintained. Redundant buildings are often more likely to be neglected as there is little incentive for the owners to maintain them, and neglect can lead to loss of fabric as weather and vandalism destroys them. Therefore heritage authorities recognise the need for managed usage changes, ensuring the continued use and maintenance of the building.

The way we live and work now can be very different to a century ago, and buildings often need to be altered or extended to make them usable. Without these changes a historic building can easily become redundant. An example of this may be reordering a historic church so that the wider community can use it when the size of congregation is too small alone to ensure its continued use and maintenance. This may involve the addition or alteration to create a disabled toilet, kitchen, better heating and lighting and possibly the removal of pews and the creation of a space inside the church which can be used for services, concerts or exhibitions etc.

Generally the test for the conservation officer on the acceptability of such an alteration will be to determine whether the alteration is justified and to ensure the impact on the historic building is minimised. It will be necessary to be able to explain how the proposed changes will promote activity in the building and how this can ensure its continuing use and related maintenance.

Redundancy of use can be a reason to convert a historic structure, and a common example is the conversion of redundant farm buildings. Old barns with roof heights and door openings that are too low are often unsuited to modern farm machinery, and farmers have often built new steel sheds to replace the old timber framed barn. Old barns are thus left to decay, but they have historic value and are often attractive features in the landscape. The conversion of these redundant buildings will ensure their continued maintenance.

Current redundancy of use is not a carte-blanche for change. Councils often ask applicants to prove that the building is redundant before they accept a change of use, since it almost certainly will change the character of the building. This is normally achieved by the owners putting the building on the market for sale or rent for a period of at least 6 months, or even a year or more. The council will want to see if anyone is interested in using it for its original use either by renting or even buying the building. Often clients assume that a building is redundant because they have no use for it, but this may not be sufficient for the planners who will want to ascertain if a user for the building can be found for its established use. They are also likely to insist that the property is advertised widely for the existing use at a realistic rent or sale price. Until this effective marketing is completed it will not be possible to submit a valid listed building application. Architects need to allow for this in the programme and advise clients of this often very onerous precondition for a change of use application.

Some uses of a redundant building will preserve more of the existing building unaltered than others, and are therefore favoured by the heritage authorities. Old agricultural buildings or industrial buildings often carry the most value when they are converted to housing, but the residential use of these previously open buildings can transform and destroy much of their historic significance. Dwellings are often cellular and people want to domesticate the space outside the building with gardens and sheds and parking. A conversion of these buildings for a commercial use such as a workshop or office may preserve more of the open character and leave more of the significant fabric untouched. Many councils have policies that ask applicants not only to prove a building is redundant but also that the applicant has actively advertised it on the open market for a more acceptable, less transformative use than residential, for a significant period. Therefore clients often need to advertise the building, both for its current use or subject to planning, for a commercial use at the same time.

Conservation officers often hold the view that an alteration which is **reversible** is more acceptable than one which is not. Explaining how the alteration is constructed so it can be removed can often assist the application. For example; where a new floor is needed at a higher level it could be built of lightweight construction such as lightweight timber joists on timber studding which can be removed leaving the original intact floor untouched below. The alternative strategy of constructing the new floor in concrete on a hard core base is far less reversible.

It should also be considered that a building which may be redundant in itself may be part of a broader context such as a group of buildings or a larger area of land. Any change to the redundant part may still affect the value of the whole. In the grounds of a listed building an owner may see an opportunity to realise a development opportunity by selling, what they consider to be redundant land, for a new building. A large listed house may have space in the garden for a new house which the current owners would be happy to sell on. The planners would consider the application on this 'redundant' land against the future of the historic building. The sale of the land may leave only a small garden and reduce the attractiveness of the house for future owners and therefore prejudice its future maintenance.

Clients often wish to extend their historic building where there is a need for extra space. But how much a building can be extended will relate to its architectural and historic significance and the size of the proposed extension in comparison to the original building. The parameters must be understood in order to advise clients effectively.

A listed historic building which was designed as the vision of one known architect which has had no extensions, and is complete, may be a very difficult to extend because any alterations would erode its architectural character.

To preserve the character of a listed building new additions are not normally allowed to dominate the historic building or mask the original built form. A historic building which has been regularly altered and extended may be much easier to obtain consent for in relation to a further extension. A small extension to a large building also would normally have less impact and therefore be more acceptable than a large addition to a small building.

An architect needs to advise clients carefully when considering extensions, as often only modest extensions will be acceptable, and these are often well below the ambitions of the client. It can be very frustrating for the owner of a listed building to realise the planners will oppose an extension, with the only alternative remaining for them is a move to a larger building. The architect may need to explain to them that their needs are not the primary concern of the planners. If the building is not under threat and other owners would take the property on unaltered, the current owners can expect little sympathy for their aspirations if they will adversely affect the historic and architecturally significant fabric.

[f] Alterations to make a building more sustainable

The government's stated aim is to improve the energy efficiency of our existing building stock. When the government signed the Kyoto Agreement it committed itself to reducing our CO_2 emissions by 20% from 1990 levels by 2010. This target has been hard to achieve partly because the burning of fossil fuels to provide energy for our buildings represents 46% of the UK's emissions, and also because most of our houses were built when energy and insulation standards were significantly lower than today. Annually we now are building so few new homes that they represent less than 1% of our national housing stock. Therefore if the Government is serious about meeting its international obligations to help prevent climate change and reach international targets in this regard, improving the energy performance of our historic buildings is going to be necessary. Indeed if our pre-1919 buildings are to be maintained we need to ensure that their owners can continue to use them and afford to light and heat them. Diminishing world resources and higher energy prices mean that finding suitable ways to reduce the running costs of older buildings will be a growing source of work for architects and surveyors in the future. However it is not always easy to achieve improvements to listed historic buildings, as the standard ways to upgrade their energy efficiency may change their character and increase rates of fabric deterioration.

The Energy Act 2011 provided a new programme of support for our existing building stock called the 'Green Deal', which intends to reduce carbon emissions cost-effectively by encouraging owners to improve the energy efficiency of existing properties. The Green Deal in theory gives a financial mechanism that eliminates the need to pay upfront for energy efficiency measures and instead provides a means to ensure the savings can be met and installation costs recovered through electricity bills. To date its uptake by home owners has been minimal and many of the improvements being sold by green renovation companies are not suitable for historic houses.

Our historic buildings also represent a considerable investment of embodied energy and finite resources which were used to make them. The embodied energy includes all the manufacture and transport of building materials and the energy used in construction. Indeed this embodied energy generally represents over 5–10 years-worth of energy used to heat and light a building. Therefore improving the energy efficiency of our existing building stock is often a more sustainable option than constructing new replacement homes.

As global resources and the costs of heating and energy become prohibitively expensive, poorly-insulated historic buildings will become less attractive for modern uses. Their future use and maintenance will be in jeopardy without sensitive modifications to ensure the future of our heritage.

There is also a requirement in Part L of the Building Regulations[3] to improve the carbon footprint of older buildings, with insulation, double glazing etc, which is discussed in more detail in Chapter 8.

Risks of thermal upgrading

Most historic buildings are constructed of porous materials which are not vapour tight, and they do not incorporate any damp proof courses or moisture barrier membranes.

Moisture which is generated in the building or makes its way through the structure and coverings is allowed to escape through the fabric. It is often said that historic buildings 'breathe'. The building regulations are largely written to control modern methods of construction which seeks to control moisture by sealing the building and making the fabric impervious to moisture and vapour, sealing gaps and adding barriers. If impervious materials including insulation and linings are introduced into historic buildings the inherent moisture which used to pass through the structure, and the moisture produced by occupants is prevented from escaping. This can lead to condensation occurring up against historic fabric, and this water can encourage decay from water damage, rot, and insect life such as wood boring beetles. Historic buildings tend to have the following characteristics:

- Their external fabric and structure will tend to be wetter than in modern buildings
- There is unlikely to be any damp proof courses or vapour membranes in the structure, nor are they needed
- They tend to be made of porous materials which need to allow moisture to evaporate both to the inside and outside. If these porous materials are not allowed to let water evaporate from them, the traditional materials tend to decay faster than usual. For example when modern impermeable paints are used on old solid walls covered in lime plaster the trapped moisture behind the paint can lead to paint peeling and plaster decaying
- Ventilation rates inside old buildings tend to need to be higher than for modern buildings in order to prevent condensation and accelerated fabric decay
- Modern heating systems fitted in traditional buildings may cause internal moisture levels to rise as evaporation from the fabric rises, and this can cause problems such as encouraging mould growth on walls where ventilation rates are low or the building is intermittently heated.

Traditional materials such as lime plasters and cob earth walling often have the ability to conduct moisture through capillary action, as well as the tendency to draw moisture out of the air by their hygroscopic action. These properties can help regulate the moisture content in a traditional building and provide a healthy and pleasant environment to occupy. Erecting vapour resistant barriers or blocking sources of ventilation can break down the conditions which keep the internal environment in balance, leading to increased humidity and fabric decay.

It is therefore very important to consider these issues carefully when upgrading a historic building in order to improve its thermal performance, especially when introducing new insulating materials into a formerly un-insulated structure. For example:

- Can insulation be added without increasing the risks of condensation?
- How much insulation can be added without increasing the risk of condensation?
- Is it possible to insulate without trapping moisture in the fabric?
- What type of insulation can be used and what are its properties? In historic buildings insulation is used without a vapour barrier, to allow a controlled amount of moisture to pass through it, without encouraging interstitial condensation. Natural or novel materials are often advocated, but care must be exercised to ensure that these materials have been adequately tested over time, or carry manufacturer-insured warranties, and of course comply with regulations. It is often wise to computer model the effects of adding

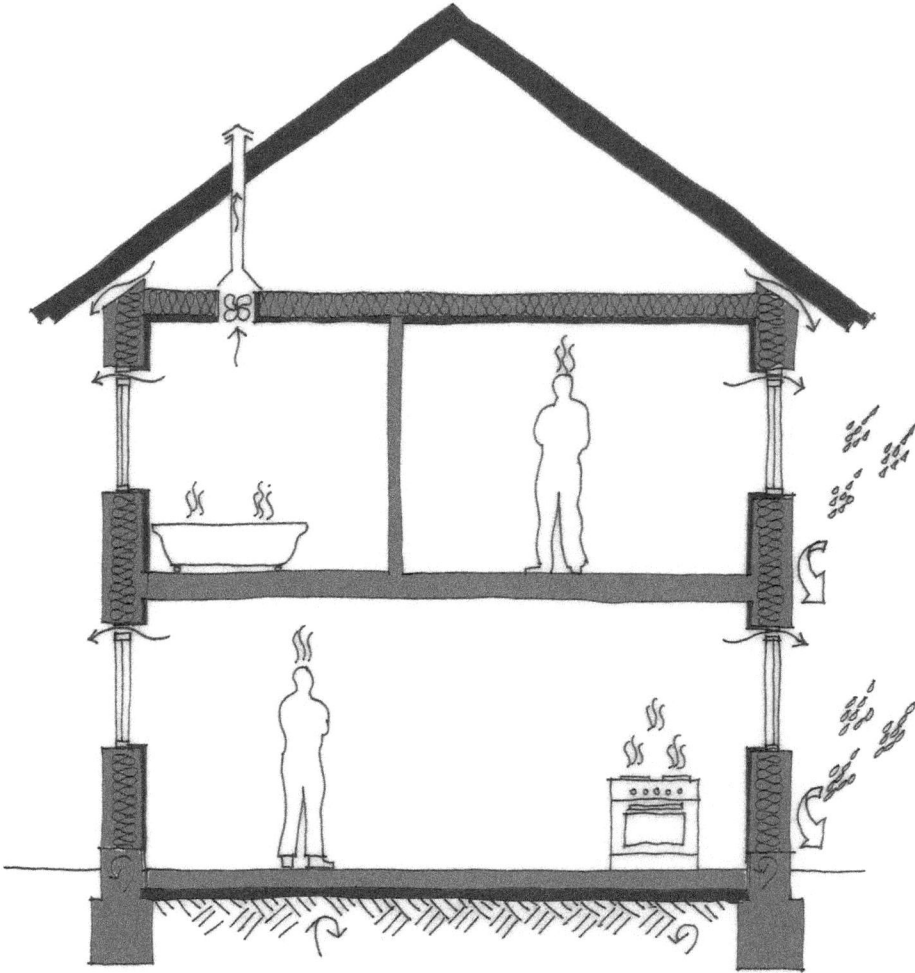

MODERN CONSTRUCTION MOISTURE CONTROL

FIGURE 3.3 Image illustrating the modern construction moisture control method.

insulation to historic fabric even when it is **breathable**, to establish the likelihood of interstitial condensation occurring in different climatic conditions. Manufacturers and suppliers of insulation materials often provide this service free or for a modest fee.

- Is it possible to install the insulation uniformly, without forming points where thermal bridging leading to condensation will occur? The incidence of condensation occurring will be higher in cold, unheated spaces and voids where they occur close to well heated spaces. Therefore even distribution of heat from a well-controlled heating system needs to be provided, or alternatively insulation between the spaces.
- Can moisture from showers, kitchens etc. be prevented from being being distributed around the building, leading to condensation where cold bridging occurs? This is normally achieved with the use of natural or mechanical ventilation systems, but this may not be acceptable in some special historic interiors.

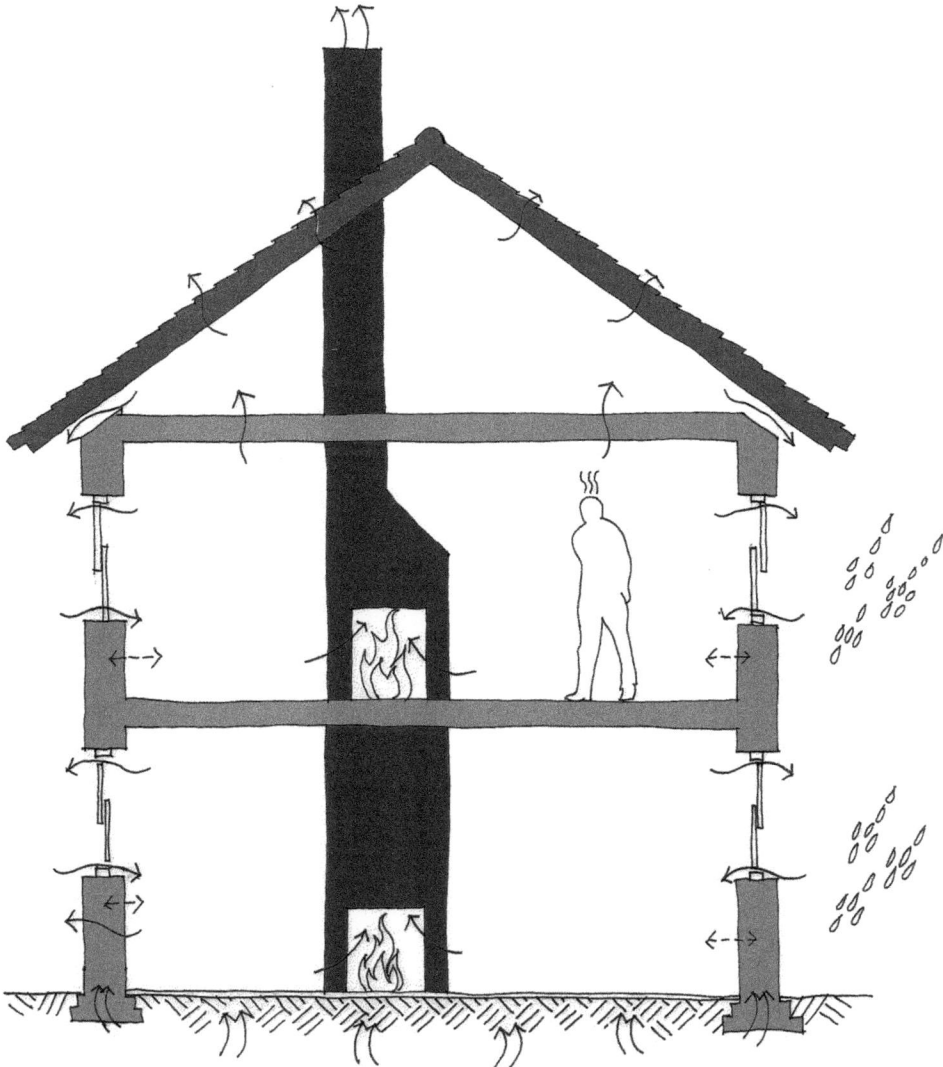

TRADITIONAL CONSTRUCTION MOISTURE CONTROL

FIGURE 3.4 Image illustrating the traditional construction moisture control method.

Continuing research

The need for better ways to improve the energy efficiency of our existing building stock is leading to a period when much research is being done and new techniques and materials developed. Historic Scotland, English Heritage and SPAB are undertaking[4] long-term research into the way historic buildings provide a comfortable environment, and indications are that our historic buildings can be surprisingly energy efficient with relatively minor improvements. Interventions such as the careful use of insulation in roof voids or dynamic heating programmers, which sense the building's conditions and ensure a room is at a certain temperature at a certain time, can be very efficient. Simple changes to control ventilation such as installing register plates – a traditional device to shut a chimney when not in use – can dramatically improve heating bills by reducing heat lost.

Making historic buildings energy efficient

How much can be done to improve a carbon footprint will be roughly proportional to the significance of the fabric. Grade I listed buildings such as Westminster Abbey may be difficult to modify sensitively, while for many Grade II listed buildings the challenge may be easier.

Simple changes to services are unlikely to require consent, such as installing new efficient lamps, upgrading to more efficient boilers, intelligent heating controls and sensors, or register plates in chimneys (to stop the loss of hot air up chimneys). However there are other alterations that are becoming more common which generally need consent and careful consideration such as:

- Increasing insulation in walls, roofs or floors. Not only will consideration need to be given to the dangers of preventing moisture escaping through new linings and insulation, but lining walls and ceilings can also be difficult to achieve without destroying or hiding significant fabric. They are often lined with panelling or historic renders, plasters, or original paint, and have original cornices or skirting, making insulation of walls internally unacceptable.
- Windows and doors are often one of the keys to the significance of a building and their character can be destroyed by double glazing and alterations to make them more efficient. It is for this reason that conservation officers resist double glazing which relies on deeper rebates and wider glazing bars than the original windows. These modern deep rebates are needed to hide the sealed edges of the double glazed units which will decay if they are exposed to sunlight. Old windows may also contain handmade types of glass, such as crown or cylinder, which is historically significant and adds character to the building.
- The addition of photo-voltaic (PV) cells, solar water panels on roofs may be acceptable where they will be hidden and have no impact on the significance of the buildings. There are now several Grade I and II churches which have had solar panels fitted on south-facing roofs which are hidden from view by parapets or are fitted in hidden valleys between buildings. Some of our historic stately homes have had PV panels fitted on hidden lead roofs. Consideration will be needed to the sensitive routing of cables and the installation of the inverters and control equipment.
- Wood chip boilers can be used to provide a sustainable heat source, but may need to be housed away from the main building. These units require supplies of wood chips and can be larger than oil or gas boilers.
- Air source heat pumps are becoming more common in historic commercial buildings and in housing since they can be easily installed and the supply pipes to the units can be very small and more easily concealed in historic buildings than conventional plumbing. However, they need to have condenser pump units housed discretely outside and can be expensive to run in uninsulated buildings.
- Ground source heat pumps can be used to heat historic buildings gently and economically. However these systems are expensive to install and normally more affordable on highly insulated modern buildings. In certain ground conditions they rely on

replenishment of the ground heat in the summer by cooling the building and pumping the heat back into the ground. This can be a problem where they serve a cool historic building such as a church where no summertime cooling is needed. Ground source heat pumps are normally fed from deep bore holes or by placing coils across a piece of open ground. The installation of these systems is normally subject to a planning permission, as it is considered to be an engineering operation. In the grounds of historic buildings it may be necessary to undertake an archaeological investigation of the site to access the risk of damaging the hidden buried heritage.

Notes

1   See SPAB website for details of their technical advice services on www.spab.org.uk For the services available for members of the Listed Property Owners Club see their web site on www.lpoc.co.uk
2   Heritage at Risk Register is available at www.risk.english-heritage.org.uk
3   HM Government (2010), *The Building Regulations: Approved Document L1B (Conservation of Fuel and Power in Existing Buildings Other than Dwellings)* and, *Approved Document L2B (Conservation of Fuel and Power in Existing Buildings Other than Dwellings)*
4   Rye, C. et al. (2011–13), *SPAB Research Interim Reports 1-3*, SPAB Publications, available at www.spab.org.uk

Further reading

> Department of Communities and Local Government, Department of Media and Sport, English Heritage (2010). *PPS5 Planning for the Historic Environment: Historic Environment Planning Practice Guide*, London, English Heritage, available from www.english-heritage.org.uk/publications. This guide is still valid and a Government endorsed document, but is expected to be superseded by a Good Practice Advice Guide now the NPPF is in place.
> Department of Communities and Local Government (2012), *National Planning Policy Framework*, available from www.gov.uk/goverment/publications
> English Heritage (2011), *Energy Efficiency and Historic Buildings – Application of Part L of the Building Regulations to Historic and Traditionally Constructed Buildings*, London, English Heritage, available from www.english-heritage.org.uk/publications

CHAPTER 4
GOVERNMENT POLICY FOR HERITAGE ASSETS

Just as vital as an understanding of the underlying principles of conservation, is an understanding of how those principles are transferred into practice. Architects who work on historic buildings need to know how legislation and government policy applies to heritage assets if they are to advise clients correctly.

In this chapter we consider government policy that exists in legislation, while in the next chapter we discuss the current designation system and legislative framework in more detail.

The government has recently radically simplified its statement of policy in the National Planning Policy Framework (NPPF)[1], condensing most of the former series of documents and statements to a single, 57-page document. The policy sets out the government's requirements for the English planning system and the framework within which councils can produce their own local plans and policies. The Framework addresses government policy for 'conserving and enhancing our historic environment', and has replaced the previous principal statement of government policy called the Historic Environment, Planning Policy Statement 5 (PPS5)[2], which together with the accompanying practice guide, explained how local authorities had previously administered the complex heritage protection system. PPS5 had itself only recently superseded two lengthy and more prescriptive guidance documents, PPG15[3] and PPG16[4], which not only outlined the workings of the legislative framework for different heritage assets, but also provided practical advice on managing change to historic buildings. This earlier guidance included practical advice about what alterations could be made to a designated heritage asset. PPS5 and the accompanying practice guide were of a more philosophical bent than their predecessors, setting down principles to be observed rather than specific methodologies. Whilst PPS5 has been cancelled, the accompanying practice guide remains extant for the time being and continues to be a relevant source of planning guidance.

The new National Planning Policy Framework has distilled government policy for the historic environment into three pages of principles to be observed and methods of dealing with applications to change historic buildings. It can be argued that the philosophy now shaping government policy for the historic environment can trace its roots to the principles of the ICOMOS convention, and can be seen to interpret them as part of a practical process which intelligently and holistically manages change within a legislative framework.

We have seen a trend over the last few years for government policy to move away from specific recommendations and practical guidance towards statements of principle. The NPPF will offer detailed guidance of good practice to be established through standards and guidance produced by professional bodies and quasi-governmental organisations. One could expect the planning system to work in a similar way to the modern building regulations, which are now a relatively short statement of government policy relying for interpretation on documents such as British Standards. A new British Standard guide to the principles of conservation of historic buildings (BS 7913: 2013) has been published which compliments the NPPF.

Policy in Northern Ireland still has guidance which follows the more prescriptive format founded upon PPG15[5], whilst the Welsh Government has published its own policies in Chapter 6 of Planning Policy Wales (Edition 6 2014)[6], which is supplemented by Conservation Principles for the Sustainable Management of the Historic Environment in Wales (2011)[7]. This Welsh policy guidance emphasises the same principles of good conservation practice promoted in the English PPS5. Scotland already has similar high-level principles promoting good conservation practice embedded in its Historic Environment Policy (2011)[8] and the more recent Scottish Planning Policy (2014)[9].

The National Planning Policy Framework (NPPF)

The NPPF clearly states that there is to be a presumption in favour of sustainable development. The ministerial foreword to the NPPF defines sustainability as ensuring better lives for the present population, without relegating worse lives for future generations. In paragraphs 12.6 heritage assets are also recognised in the Framework as an irreplaceable resource that need to be conserved for future generations in a manner appropriate to their significance.

Section 12 of the NPPF sets out the principles and policies that councils should follow for the conservation and enjoyment of the historic environment, and which should be taken into account when assessing applications to change the historic environment. There is no detailed practical guidance given for local councils, owners and applicants to interpret and implement these national policies, making it even more vital that architects and surveyors understand the commonly accepted principles of good conservation, such as those set down in the ICOMOS guidelines, before embarking on changing a historic asset.

The NPPF explains the government's objectives for the management of the historic environment, and is largely there to guide the system operated by local planning authorities. In paragraph 126 it charges individual authorities to develop a positive strategy for the conservation and enjoyment of the historic environment. In future, it can be expected that different planning authorities will develop policies for their own specific local environment within their own local plans. Thus it will be essential for architects and surveyors to familiarise themselves with local planning guidance as well as the national policy guidance. These changes can be seen as part of the current government's localism agenda. The NPPF also emphasises the desirability of new development making a positive contribution to the local character and distinctiveness of a place, which is normally derived from historic buildings and landscapes.

The concept of significance within the NPPF

Paragraph 128 of the NPPF requires any applications for planning, listed building consent etc., to provide information on the significance of the heritage asset and its contribution to its setting.

The need to evaluate and understand the significance of any particular heritage asset is key to successfully applying the guidance in NPPF, just as it is seen as central to following the ICOMOS Charter. Significance is defined in the NPPF in the Glossary as *'the value of the heritage asset to this and future generations because of its heritage interest. That interest may be archaeological, architectural, artistic, or historical. Significance derives not only from a heritage asset's physical presence, but also its setting'*[10]. It is therefore similar to the ICOMOS Burra Charter's definition discussed in Chapter 2. The setting of the heritage asset is also clarified in the Glossary as *'the immediate and extended environment of a place that is part of or contributes to its cultural significance and distinctive character'*[11].

The scope of the NPPF – what defines a Heritage Asset?

The definition of a Heritage Asset in the NPPF is wide and covers 'a building, monument, site, place, area or landscape identified as having a degree of significance meriting consideration in planning decisions, because of heritage interest'[12]. Paragraph 128 requires the significance of all heritage assets to be assessed, and treats the whole spectrum of our national heritage assets in a similar way, be it an archaeological site, a historic battlefield, a historic shipwreck or a listed building and a conservation area. The level of detail of the assessment should vary only in proportion to the asset's importance. In addition to these designated assets, the significance of any non-designated heritage asset identified by the local authority on a **local list** will also need analysis. This policy will reinforce the status of the local lists of historic assets, which some local planning authorities have prepared for buildings which they consider important locally, but do not meet the threshold of a national designation to make them listed buildings. Local planning authorities also often require non-designated historic buildings in conservation areas to be treated as heritage assets requiring analysis of their significance.

Archaeological interest

If there is the possibility of items of archaeological interest being present in the development area, even if they are unscheduled, the local planning authority can require a desktop archaeological assessment prior to validating and determining an application, and/or impose conditions on the planning consent to record or avoid damaging unrecorded archaeology. This is a consequence of paragraph 128. In the NPPF Glossary, an asset is defined as having archaeological interest if it holds, or potentially holds, 'evidence of past human activity worthy of expert investigation at some point. Heritage assets with archaeological interest are a primary source of evidence about the substance and evolution of places, and of the people and cultures that made them'[13].

The broader setting

It is important to realise that in paragraph 129 there is now a duty on the local planning authority to identify and analyse the significance of any heritage asset which may be affected by a development proposal. They will therefore need to consider the setting of the building to see if that is affected by the new proposal. For example, if a new development is to be sited alongside a listed historic building, the architect needs to consider whether this will detract from the historic building's setting.

Architects need to be able to analyse significance

Councils have therefore been instructed by government policy to consider the archaeological, historic, and architectural significance of any application on historic buildings, and this has now become a duty for all applicants. The analysis of heritage value has thus been brought firmly into the heart of a large percentage of planning applications be they for designated historic assets, those on local lists or unlisted historic buildings in conservation areas. The definition of a heritage asset is so wide, that architects should expect to be asked to evaluate the significance of any historic structures when making an application if the local planners demand it. Architects will now need to have the required skills to be able to evaluate the heritage significance of any existing building(s) they are working on. It is becoming a professional duty to seek to balance the needs of clients with the value of the heritage asset to ensure that important, irreplaceable historic fabric is not lost to future generations without good reason. Given that architect's training covers the history of architecture, understanding of building technologies and the way society uses buildings, architects should be the most competent of all building and construction professionals to evaluate the significance of buildings. Accredited conservation architects have to demonstrate their skills in analysing a building's significance to be included on the Register, and their expertise is a vital part of the design team when working on complex historic buildings. However on simpler historic buildings most architects should be able to develop these skills. In Chapter 5 the process of analysing a building's significance is discussed in more detail.

Proportionality – how much detail is needed?

When making a planning or listed building application, the NPPF will evaluate the strength of the evidence concerning whether the significance of the building is proportionate to the value of the heritage asset. Paragraph 128 says 'the level of detail should be proportionate to the assets' importance and no more than is sufficient to understand the potential impact of the proposal on their significance'[14]. In practice this means when working on a Grade II listed domestic vernacular property where little is recorded, the architect is only required to provide a small **heritage statement**, detailing what is known about the building and what can be deduced from site analysis. However where a large change to a more significant Grade I listed building is proposed, a very detailed heritage statement would be appropriate. The principle of proportionality will also be applied when considering an application to change listed property. This is emphasised in paragraph 132 where it is noted 'that substantial harm to, or loss of a Grade II listed building, park or garden should be exceptional'[15]. It goes on to explain that 'substantial harm to or loss of the designated asset of the highest significance, notably scheduled monuments, protected wreck sites, battlefields, grade I or II* listed buildings, Grade I or II* registered parks and gardens and World Heritage Sites should be wholly exceptional'[16]. The loss of, or a significant change to a designated heritage asset, will need clear and convincing justification proportionate to its value and designation.

Balancing harm and need

It is important to realise that government policy holds a presumption in favour of the conservation of a heritage asset as it is, unless there is a clear need for change. Therefore when making an application to change a heritage asset, justification why the change will be necessary is the first point of reference. The heritage authorities will then consider the balance of harm to the significance of the structure against the needs of the client. This is discussed in more detail in Chapter 6.

Levels of harm

In paragraphs 133 and 134 of the NPPF document, two levels of tests are given to local planning authorities for their assessment of applications in order to consider the balance of harm and needs; one level is where there will be substantial harm or total loss of significant fabric, and the second level is where there will be less than substantial harm. Paragraph 133 explains that where there is substantial harm to or total loss of significance of a designated **heritage asset**, local planning authorities should refuse consent unless a substantial public benefit that outweighs the loss can be demonstrated, and all of the following apply:

• the nature of the heritage asset prevents all reasonable use of the site;
• no viable use of the heritage asset itself can be found in the medium term that will enable its conservation;
• conservation through grant funding or some form of charitable or public ownership is not possible;
• the harm to or loss is outweighed by the benefits of bringing the site back into use.

These are evidently stringent tests, which often take time and detailed analysis to satisfy, and in consequence it is rare for substantial harm to be allowed to a listed building. The test for impacts which are less than substantial is far less onerous. Paragraph 134 requires 'where a development proposal will lead to less substantial harm to the significance of the designated heritage asset, this harm should be weighed against the public benefits of the proposal, including securing its optimum viable use'[8].

There is no definition of what 'substantial' means within the NPPF. Sometimes it is self-evident when 'substantial harm' is occurring, for example, where a listed building is proposed to be demolished. However the definition of 'substantial' is clearly going to be a matter of much debate in the future, and is highly likely to be argued in future appeals against planning refusals. It can be assumed that the principle of proportionality recognised in the government guidance should be applied. A minor change to a Grade I listed building which has had few changes over the centuries, and was built by a single famous architect, may be seen as 'a substantial change'. Whereas the removal of a whole wall in a large listed vernacular building, similar to many other listed examples, which has seen many changes over its existence, might be argued, as being 'less than substantial'.

Sustainability and the NPPF

The NPPF confirms the Government's policies and the key role of the planning process in shaping our environment to reduce greenhouse gas emissions, to minimise vulnerability and provide resilience to climate change, as well as supporting the delivery of renewable energy sources. Unlike in PPS5, the previous statement of government policy for historic buildings, there is no specific mention of the effects of climate change on historic buildings. However NPPF does, at paragraph 131, emphasise the 'desirability of sustaining and enhancing the significance of heritage assets and putting them to viable uses consistent with their conservation'[18].

The rising cost of energy is putting the use of many uninsulated historic buildings in jeopardy. There is a growing demand for architects to develop acceptable ways of making our historic fabric more sustainable without destroying their heritage value. The emphasis in the government's new policy is on preserving heritage assets, but there are clearly public benefits which are encouraged by the NPPF to improve the energy efficiency of our historic buildings. However, where conflicts between climate change objectives and the conservation of heritage assets are unavoidable, local planning authorities will need to weigh up the public benefit of mitigating the effects of climate change against the significance of the heritage asset. There are also clear public benefits and sustainable advantages in keeping heritage buildings in use, avoiding the consumption of building materials and energy, as well as the generation of waste from the construction of replacement buildings.

Recording

Paragraph 141 of the NPPF document imposes duties on Local Planning Authorities to record any evidence of our historic past which becomes available as part of their plan making or development management process. This is normally done by noting the location of records into the relevant **historic environment** record. In England, these records can normally be accessed by the general public via English Heritage's Heritage Gateway website, which gives the location and nature of the record[19]. When the loss of the whole or a significant part of the heritage asset is justified and a consent given, paragraph 141 also requires the developer to prepare an appropriate record of the fabric being lost. Such record will probably take the form of architectural drawings and photographs, or measured archaeological surveys and reports. The records often need to be submitted to the local authority to discharge a planning condition. The council will then lodge the recording in an appropriate location. Sometimes clients are required to lodge these records directly with regional or county archives, or the National Archives.

Ecclesiastical buildings

Some listed buildings which are used as places of worship specifically by the Church of England, the Roman Catholic Church, the Methodist Church, the Baptist Union of Great Britain and the United Reformed Church, do not require Listed Building Consents, but require a consent from bodies set up by their churches. This system is known as the **Ecclesiastical Exemption** and is explained in more detail in Chapter 7.

The existing government guidance for the operation of the Ecclesiastical Exemption will probably be brought into line with the NPPF. The guidance was recently updated to bring it into line with the policy guidance in PPS5, but was only issued as draft guidance pending the NPPF.

The fundamental principles espoused in NPPF all apply to these church buildings in the same way as with any other heritage asset, namely the consideration of the significance of the building, the proportionality of the evidence needed, the proportional response to an application depending on its significance, and the balancing of needs against harm. Draft guidance has been issued by the DCMS (Department for Culture Media and Sport) but its adoption is pending.

Notes

1   Department of Communities and Local Government (2012), *National Planning Policy Framework*, available from www.gov.uk/goverment/publications
2   Department of Communities and Local Government (2010), Planning Policy Statement 5
3   HM Government (1994), *Planning Policy Guidance 15, Planning and the Historic Environment*
4   HM Government (1990), *Planning Policy Guidance 16, Archaeology and Planning*
5   Planning Service (DOENI), (1999), *PPS 6: Planning, Archaeology and the Built Heritage*
6   Welsh Assembly Government (2014) *Planning Policy Wales*, Chapter 6, Conserving the Historic Environment, available at www.wales.gov.org
7   Cadw, Welsh Assembly Government (2011), *Conservation Principles for the sustainable management of the historic environment in Wales*, available at www.cadw.wales.gov.uk
8   Historic Scotland (2011), *Scottish Historic Environment Policy*, available from www.historic-scotland.gov.uk
9   The Scottish Government (2014), *Scottish Planning Policy*, available from www.scotland.gov.uk
10 Department for Communities and Local Government (2006), *National Planning Policy Framework*, p. 56
11 Australia ICOMOS (2013), *The Burra Charter: The ICOMOS Charter of Places of Cultural Significance: Article 1.1.2*, p. 3
12 Department for Communities and Local Government (2006), *National Planning Policy Framework*, p. 52
13 Department for Communities and Local Government (2006), *National Planning Policy Framework*, p. 50
14 Department for Communities and Local Government (2006), *National Planning Policy Framework*, para 128, p. 30
15 Department for Communities and Local Government (2006), *National Planning Policy Framework*, para 132, p. 31
16 Department for Communities and Local Government (2006), *National Planning Policy Framework*, para 132, p. 31
17 Department for Communities and Local Government (2006), *National Planning Policy Framework*, para 134, p. 31
18 Department for Communities and Local Government (2006), *National Planning Policy Framework*, para 131, p. 31
19 See www.heritagegateway.org.uk/gateway

Further reading

> Department for Communities and Local Government (2006), *National Planning Policy Framework*, available from www.gov.uk
> Department of Communities and Local Government, Department of Media and Sport, English Heritage (2010), *PPS5 Planning for the Historic Environment: Historic Environment Planning Practice Guide*, London, English Heritage available from www.english-heritage.org.uk/publications
> Department for Communities and Local Government (2010), *The Operation of the Ecclesiastical Exemption and related planning matters for places of worship in England*, available from www.gov.uk
> Morrice, R. (2009), *Ecclesiastical Exemption – Historic Churches and Building Conservation Directory 2009*, Tidsbury, Cathedral Communications Ltd, also available at www.buildingconservation.com

CHAPTER 5
HERITAGE DESIGNATION AND THE LEGISLATIVE FRAMEWORK

This chapter looks at the how heritage assets are designated, the legislation which protects them and the consents or permissions needed to carry out any building work.

Just as there are many different types of heritage asset, there are many different forms of protection specified by the law. All seek to control whether heritage assets can be adapted, changed or demolished. The different forms of designation used are:

- Scheduled monuments and areas of archaeological importance
- Listed buildings
- Buildings of local interest
- Conservation areas
- Historic parks and gardens
- Historic battlefields
- World heritage sites.

In addition to these designations there are heritage assets which are locally designated or acknowledged less formally, they include:

- Local lists
- Non-designated heritage assets

There are proposals however to amend the process of designation. In 2007 the Department for Media and Sport and the Welsh Government published a joint white paper, *Heritage Protection for the 21st Century*[1],which proposed the unification of the many designation systems and consent regimes into one national register of heritage assets. This reform was not carried out, but may be revived again now the National Planning Policy Framework has been adopted, and when the parliamentary legislative timetable allows. With so many different designations of heritage assets, it is important to understand the differences between them and the scope of their protection.

Monuments

In the late-Victorian period the lobbying of concerned conservationists, including William Morris, led to the first piece of legislation protecting our heritage in the form of the Ancient Monuments Protection Act 1892. This Act protected 69 of Britain's most famous ancient historic sites, such as Stonehenge. All the sites were prehistoric rather than occupied buildings. This act was revised in 1913 when the definition of a monument was widened to include built form, 'the preservation of which was in the public interest'[2] by virtue of their archaeological, artistic, historic, or architectural interest. The definition specifically excluded ecclesiastical buildings which were in use.

The Act was further revised in 1953, and was superseded later by the Ancient Monument and Archaeological Areas Act of 1979[3], which introduced the concept of 'areas of archaeological importance'. Almost a century after the first legislative protection of monuments, the current legislation, the National Heritage Acts came into force in the 1980's[4]. Under this legislation English Heritage in England, Cadw in Wales and Historic Scotland, (on behalf of their respective ministries), maintain lists of 'scheduled ancient monuments and archaeological sites'. In Northern Ireland there is similar legislation, however monuments and archaeological sites are more simply known as 'Historic Monuments'. There are now around 18,300 scheduled monuments in England and Wales, and around 1700 in Northern Ireland and 8000 in Scotland. The numbers of scheduled monuments is growing, and English Heritage estimate there may be as many as 60,000 sites which deserve inclusion. The definition of what constitutes a scheduled monument or archaeological site is broad and can include the following:

• Any scheduled ancient buildings above or below the ground
• A cave of archaeological interest
• Any site containing the remains of an ancient building, structure, excavation or work of any kind
• Any site containing remains of machinery, a vehicle, vessel or aircraft which is of particular historical significance

Scheduled monuments need to be historically important, and range from Bronze Age earth mounds to the remnants of relatively modern industrial activity which are rare and of historic significance, such as disused gunpowder mills from the late-nineteenth century. The legislation specifically excludes ecclesiastical buildings, although the graveyards around some churches are occasionally scheduled where they contain significant archaeology, such as the archaeological remains of very early ecclesiastical buildings.

There can also be significant archaeology in rivers and along the coastline, such as historical remnants of harbours, vessels and defensive structures, and these can also be scheduled as ancient monuments.

Scheduled monument consent

Scheduled monuments and archaeological sites are clearly very vulnerable to being damaged by construction work, and therefore scheduled monument consent will be needed to carry out virtually any work in these areas. This includes maintenance, any digging into the ground, or even gardening activities such as the removal of trees or shrubs. The need for scheduled monument consent does not override the requirements under other planning legislation, such as planning or listed building consent. It is a criminal offence to carry out any work to a scheduled monument or site without consent.

Scheduled monument consent is applied for directly from the relevant National Heritage body; English Heritage (via the DCMS), Cadw, Historic Scotland or the Northern Ireland Environment and Heritage Service. When dealing with scheduled monuments, architects will normally need the advice of consultant archaeologists to guide them through the process of evaluation, mitigation and application.

Areas of archaeological importance

In addition to the scheduled monuments legislation, five of our most historic city centres have special protection as areas of archaeological importance. These areas were designated in 1984 and are Canterbury, Chester, Exeter, Hereford, and York. In these areas the local council's archaeological officer will need to be consulted and agree to any groundwork to determine whether the site has any archaeological implications. Before planning consent is issued, developers will probably be asked to submit a desktop assessment that should address the issues of the potential archaeological impact the development proposals on heritage assets might have, and how this impact may be mitigated. If desk-top research is insufficient to properly assess the effects of the proposal, then a field evaluation by an archaeologist may be required.

Operations which disturb or clear ground in areas of archaeological importance will need to give six weeks notice to the council's archaeological officer before starting work. The notice to be used can normally be found on the relevant council website. Consent may well be needed for works as small as laying a new drain, or providing statutory services to a building, as well as constructing pits for large trees and shrubs.

It is a criminal offence to carry out any work in an area of archaeological importance without consent.

Listed buildings

Legislation to protect historic buildings evolved through a series of acts controlling town planning. It was not until the 1944 Town and Country Planning Act however, that there was effective control of change to historic buildings. This act allowed for a Preservation Order to be made prohibiting alteration or extension to a building. In 1947 a new Town and Country Planning Act was passed, which came into force on July 1 1948, giving the new local council planning authorities the power to make 'building preservation orders' to protect historic buildings. The act also required the Secretary of State to make a list of buildings of special architectural or historic interest to guide local authorities in their choice of buildings requiring preservation orders. This was further underlined in the Planning (Listed Buildings and Conservation Areas) Act of 1990, which continued the duty on the Secretary for State in England and Wales to compile a list of buildings of 'special architectural and historic interest'.

The first schedule of listed buildings had been hurriedly created during World War II, which consisted of buildings so important that the government of the day felt they should be reconstructed in the event of bomb damage.

In the late 1940s and 1950s surveys were carried out throughout Britain, and further buildings were added to the list of buildings which were of special architectural and historic importance. In the 1970s and 1980s there were outcries at the loss of some significant, but relatively modern buildings, including the Firestone factory, which was unscrupulously demolished during the 1980 August Bank Holiday weekend, just days before a preservation order could be placed on it. Following these incidents the government ordered a new review of buildings, significantly increasing the list, which has now grown to list around 374,000 entries which cover some half a million buildings (an entry can include more than one building in cases such as terraces). This represents about 1 in 50 buildings in the country, or 2% of Britain's building stock.

The criteria for listing

There are no size restrictions for listed buildings; there are listed beach chalets, flights of steps, walls, gravestones, pillar boxes, street lamps and telephone boxes. Buildings which are in very poor condition can be listed if they are significant, as the state of repair is not deemed a relevant consideration for not listing.

The principal criteria for a structure to be listed are as follows:

- The building is of **special architectural interest** because of its architectural design, decoration or craftsmanship; important examples of particular building types and techniques as well as significant plan forms may also become listed. The principal works of notable architects are also likely to be of special architectural interest and therefore likely to be listed.
- **Aesthetic merits:** Historic buildings which are particularly appealing in appearance can be listed, but often buildings of little aesthetic merit are also listed where they are significant historically or in terms of social and economic history.
- The building is of **historic interest**, which can be defined as relating to people or events of sufficient impact to have been written about. This includes buildings which illustrate important aspects of the nation's military, social, economic and cultural history. This should not be confused with what is often referred to as 'historic fabric', which may be old but not necessarily of historic interest.
- The building has **close associations** with a nationally important person or event.
- The building is **part of a group, which show an important architectural or historic unity,** for example a fine example of planning of squares, terraces, model villages etc. In these cases a building may not be particularly significant individually, but may still be listed if it forms part of a group that is more significant; for example, all the buildings composing a Georgian square.
- **Selectivity:** where a large number of buildings of a similar type survive, the policy of English Heritage is only to list those which are the most representative or significant examples of the building type.
- **National interest:** significant or distinctive regional buildings can also merit listing as examples of a significant local industry.
- **Age and rarity:** As older buildings are rarer, they are therefore more likely to be listed.

The survey of listed buildings, in theory, should have identified most structures worthy of being listed, however buildings are still being regularly added to the list. The age of the building has an impact on how rare it is and therefore its significance. The likelihood that a building be listed in England and Wales reflects the period when it was built as follows:

- **Before 1700:** virtually all buildings which survive in anything like the original condition from this period, should now be listed. When submitting a planning application on a property that has significant surviving fabric from before the eighteenth century there is a risk of it being spot listed. The building will almost certainly be treated as listed if there is knowledge of the early fabric.
- **1700 to 1840:** for this period most buildings will be listed, but there is some selection process.
- **1840 to 1914:** the best examples of Victorian architecture are normally listed, as well as selected particular building types which are fine examples of the social and economic history of the period. Because more buildings survive from the Victorian period the list can be more selective.
- **After 1914:** only outstanding buildings or buildings under threat tend to be listed.
- **less than 30 years old:** are only rarely listed, and generally have to be of outstanding quality and significance, and also under some threat.

Buildings which are under ten years old will not be listed. However there are now around 700 post-1945 buildings listed, including Richard Rogers' Lloyds Building in London.

The criteria in Scotland for listing are slightly different and are as follows:

- **Before 1840:** all buildings that have any quality, even if plain, and survive in anything like their original form
- **1840 to 1914:** buildings that are of the finest quality and character, either individually or as a group
- **1914 to 1945:** buildings which are good examples of the work of an important architect, or a particular style
- **After 1945:** buildings of outstanding quality and of some vintage (a very high degree of selection is exercised)

In Northern Ireland there are similar criteria for listing, and virtually all buildings dating from the nineteenth century or earlier, surviving in anything like their original form will qualify for listing. Virtually nothing is listed if it is less than 30 years old.

The Department for Media Culture and Sport have compiled a 'Publication Principles of Selection for Listing Buildings', which can be found on their website, as well as the English Heritage website: www.english-heritage.org.uk

Listing process

Although a building may not be listed, it does not mean it is not significant. Omissions were made when the listing of buildings started following the Town and Country Planning Act 1947. The listing process recognised historic buildings such as churches and older houses, but the English Heritage Inspectors did not obviously enter every building in the land, and therefore many modest properties which look younger from the outside than they are, were thus omitted from the listing process. These gaps are constantly being plugged as attitudes to significance change.

To 'list' a building, English Heritage inspectors will first carry out a desktop study and research on the building, and will then follow-up on this by visiting it and noting its features to assess its significance. Their research will compare it to other examples of the same type and age of building. They will write a report which is then considered, and a judgement about listing is made after consultations. Finally it will be the Secretary of State who must agree to it being added to the National Heritage List for England.

Applying for a building to be added to the National Heritage list

Nomination of a building to be included in the National Heritage List can be achieved by using the simple web-based form available on the English Heritage website (list.english-heritage.org.uk), or by applying to the English Heritage Protection Unit. The application can be made by anyone, not only the owner of the building. It helps with this process if the person seeking the listing can provide any history of the building, drawings, photographs and the reasons why it should be considered of special architectural or historic significance. Priority to be listed is given to buildings with significance which are under threat, be it from alteration, demolition or neglect. The application process normally takes around six months, but can take longer, or can be accelerated due to inspectors prioritising buildings under the greatest threat first.

Listed building grades

In England and Wales there are three designations of listings, Grade I, Grade II* and Grade II, while similar systems are used in Scotland and Wales. The three levels reflect the importance and significance of the fabric.

- **Grade I** listed buildings are of national and international importance (2.5% of all listed buildings in 2010). Examples of such listing are St Paul's Cathedral, the Houses of Parliament, and Buckingham Palace. Grade I listed buildings however may also be much humbler, representing rare examples of surviving structures from a period, such as a small Norman church, or they may be less architecturally interesting but significant historically such as Shakespeare's birthplace.
- **Grade II*** buildings are particularly important buildings of more than special interest (5.5% in 2010).
- **Grade II** listed buildings are of interest and warrant every effort to preserve them. (92% in 2010).

Churches form a significant percentage of listed buildings with over 14,500 listed places of worship, 4,000 Grade I, 4,500 Grade II* and 6,000 Grade II. Confusingly, some churches are still listed as being grade A, B or C referring to an older system once used by the Anglican church that is equivalent to grade I, II* and II respectively. There was also a non-statutory form of listing before 1977 which noted some buildings as Grade III but this designation was removed.

Listing in Northern Ireland

The heritage listing of buildings came later to Northern Ireland than to England and Wales. The first survey in Northern Ireland took place between 1974 and 1994. As in England, three categories are used that are basically equivalent to the Grades I, II* and II:

- **Grade A** buildings of greatest importance to Northern Ireland, including both outstanding architectural set-pieces and the least altered examples of each representative style, period and type.
- **Grade B+** buildings which might have merited Grade A status but for detracting features such as an incomplete design, lower quality additions or alterations. Also included in this Grade are buildings that because of exceptional features, interiors or environmental qualities are clearly above the general standard set by Grade B buildings. A building may merit listing as Grade B+ where its historic importance is greater than a similar building listed as Grade B.
- **Grade B** buildings of local importance and good examples of a particular period or style. A degree of alteration or imperfection of design may be acceptable. Grade B is further separated into B1 and B2 depending on a building's quality.

Currently there are around 8,500 listed buildings in Northern Ireland, which similar in proportion to England comprises around 2% of building stock. 200 buildings in Northern Ireland are listed at Grade A, 400 at Grade B+ and the remainder are Grade B.

Listing in Scotland

Scotland has a different set of definitions for its 47,700 listed buildings, which also mirrors the three tier English system:

- **Category A** buildings of national or international importance, either architectural or historic, or fine little-altered examples of some particular period, style or building type.
- **Category B** buildings of regional or more than local importance, or major examples of some particular period, style or building type which may have been altered.
- **Category C** buildings of local importance, lesser examples of any period, style, or building type, as originally constructed or moderately altered, as well as simple traditional buildings which group well with others in categories A and B.

Revisions to listing

In very rare cases listed buildings are downgraded, but it is also not so unusual for buildings to be upgraded in light of new information becoming available about their significance.

In determining whether a building is listed, each local planning authority is required to keep a public display copy of the National Heritage List available to be inspected at their offices. It is also very easy to check if the building is listed by carrying out a search on the English Heritage website (list.english-heritage.org.uk) or the English Heritage Gateway website (www.heritagegateway.org.uk).

Spot listing

Often buildings that were missed by the listing process in an area only become listed once a planning officer has visited them following the submission of a planning application. If the visit reveals a historically significant fabric that has never been recorded, this may initiate the listing process. Neighbours or local amenity groups may have concerns and realise that a locally significant building may be lost through demolition or alteration, and seek to prevent a change by lobbying the Secretary of State for Culture Media and Sport to spot list the building. In these cases the local authority can issue the owner with a temporary Building Preservation Notice (BPN). The BPN is a temporary listing of the structure, which is served by the local planning authority under section III of the Planning (Listed Buildings and Conservation Areas) Act 1990[5]. The Secretary of State will then have six months to decide whether to list the building following the issuing of the notice.

The issuing of a BPN is only done when there is a risk of loss of historic fabric through demolition or alteration should the notice not be served. Court cases have established that the notice should be issued early in the planning application period to give the client a chance to modify his/her proposals or challenge the listing. As an architect seeking to alter an unlisted, but clearly historic building, it would be wise to advise the client of these risks as it can severely delay the planning process if such notice is served.

There have been successful claims for compensation from local authorities by aggrieved planning applicants where a BPN has been issued without good cause and the building has not been listed; this can make local planning authorities cautious about using this power unless they are convinced the building is likely to become listed.

Immunity from listing

If there is a concern that a building may become listed whilst the authority considers the planning application, there is a mechanism for applying to the Secretary of State prior to making the planning application for a Certificate of Immunity (COI) for the building from being listed. This will establish whether the building is likely to be listed, and if not, it will prevent anyone trying to get it listed for a further period of five years. To obtain a COI the architect needs to apply to English Heritage using the same website as would be used to nominate a heritage asset, but stating that a Certificate of Immunity is being applied for, as opposed to nominating the asset for protection. Guidance is available on the English Heritage pages on listed buildings at http://list.english-heritage.org.uk.

The extent of listed building control

Any physical changes which alter the character or special architectural or historic interest of a listed building, or objects covered by the list's description – including those in the curtilage of the building – will need listed building consent from the local planning authority, regardless of whether these changes involve development for which express planning permission is required.

It is important to understand that the listing protection applies to the whole fabric of the building, and furthermore extends to the interior, fixtures and fittings, no matter what the grade of the listed building. It is equally important to realise that if some features are more architecturally and historically significant than others, the less significant features are still protected and consent will be needed to change them.

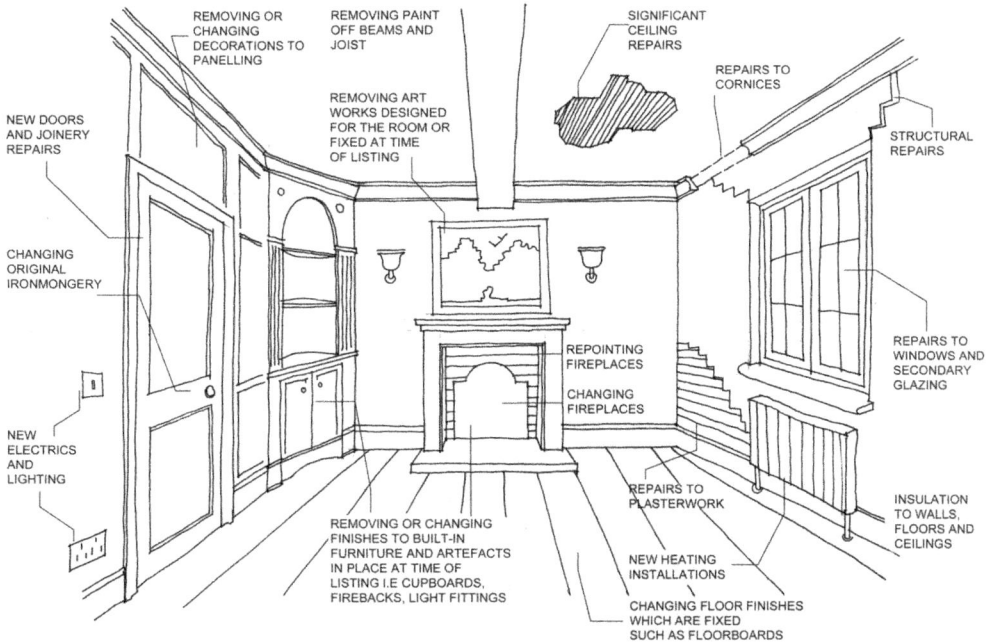

FIGURE 5.1 This sketch illustrates the type of alterations which are likely to need consent to all grades of listed buildings.

The exact extent of protection provided by listing a building on the National Heritage List has been the subject to many court cases and much debate. A Tudor house may be listed as being of special architectural and historical interest, but does this protection extend to the furnishings in the house, the sanitary ware, or the stable block in the garden?

It is obvious that a listed building derives its character not only from its external appearance, but also the internal fixings and furnishings. The doors, fireplaces, skirting boards, floorboards, staircases, door handles, ironmongery etc. all these fixed objects contribute to the overall ambience of a building, and are controlled under the legislation. However the building will also contain loose furniture, pictures hanging on walls, standard lamps etc. which also contribute to a lesser degree, to the building's character which would not usually be controlled.

The current act of parliament defines the extent of the listed building beyond the principle structure as:

- **The principal listed building itself including any extensions**.
- **The interior of the building** is covered by the legislation no matter the designation I, II*, or II. This includes finishes to floors (though not normally carpets), ceilings, staircases, fireplaces, doors and door casings, ironmongery and even early light fittings, switches and sockets.
- **Any object or structure fixed to the building at the date of listing**; this could include fixed pictures, fixed cupboards etc. As a rough guide to what interior objects might need consent to be removed, imagine the property turned upside down and shaken vigorously; any object which would fall out such as rugs, china, linen, paintings attached by

wires, loose furniture probably do not need consent; and anything which remains in place will need permission to be removed. If the interior however, including its furniture, was designed by a significant architect or designer, or is part of an interior design scheme of significance and contribute to why the building is listed, then the loose furniture, paintings, tapestries etc. despite being loose, will also need consent to be removed.

Given that many historic objects in listed houses have great value, there have been many appeals and even court cases to establish if an object is 'fixed' to a listed building. These ambiguous cases include items such as panelling, tapestries or art made specifically for the house, sculptures, garden ornaments, and bells, mirrors, and built in clocks. One important test in the current interpretation of the law includes considering whether the object was placed in the room in order to show the object off, or placed there as part of a scheme to create a planned beautiful space. Recent decisions of the Secretary of State have ruled against the removal of loose objects which would unquestionably affect the character of the building, such as tapestries made for the building, wall lights and mirrors.

If it is unclear whether the object concerned is covered by the listed building legislation it is worth discussing it in detail with a specialist accredited conservation architect or the conservation officer.

- **Any structure fixed to the principal building.** This would include modest structures such as log stores or porches where they are attached to the building even if they are of no architectural value.

- **Any object or structure and building within the curtilage** of the main listed building which at the date of listing (or possibly January 1st 1969 for buildings listed before this date), although not fixed to the building, forms part of the land and has done since before July 1st 1948. This is done because changes to these external structures may affect the surroundings in which the heritage asset is experienced and could reduce the significance of the building by destroying its context and aesthetic appeal. The objects or structures also may be of historic significance in their own right, but not of sufficient significance to have a separate listing.

Curtilage of a listed building

The extent of the curtilage can be difficult to assess and is generally defined as the land around the listed building which has historically been used in conjunction with the principal building at the date of listing. However, there is no exact legal definition of what constitutes the curtilage of a listed building, therefore the principal sources of guidance on this point have been left to legal cases and appeal decisions.

Each case needs to be assessed on its own merits. The curtilage can often be extensive or quite small, and is normally assessed by looking at historical evidence, ownership records, the physical plan etc. It may include other buildings, and the current ownership may not be relevant to the definition of the curtilage to the planning authorities. For example a listed farmhouse which was historically used with, and at the time of listing, a set of adjoining barns as a working farm yard will probably have a curtilage which includes the barns, even if they have been converted to a different use and are now are in separate ownership. This can mean that listed building consent may be required for a property which at first sight is

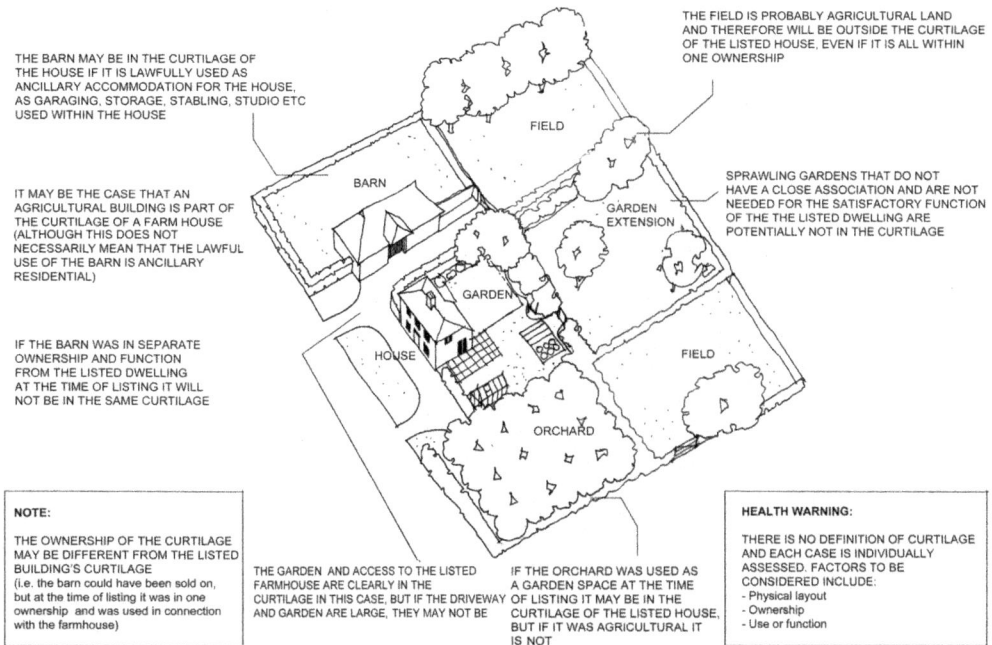

THE FIELD IS PROBABLY AGRICULTURAL LAND AND THEREFORE WILL BE OUTSIDE THE CURTILAGE OF THE LISTED HOUSE, EVEN IF IT IS ALL WITHIN ONE OWNERSHIP

THE BARN MAY BE IN THE CURTILAGE OF THE HOUSE IF IT IS LAWFULLY USED AS ANCILLARY ACCOMMODATION FOR THE HOUSE, AS GARAGING, STORAGE, STABLING, STUDIO ETC USED WITHIN THE HOUSE

FIELD

SPRAWLING GARDENS THAT DO NOT HAVE A CLOSE ASSOCIATION AND ARE NOT NEEDED FOR THE SATISFACTORY FUNCTION OF THE THE LISTED DWELLING ARE POTENTIALLY NOT IN THE CURTILAGE

IT MAY BE THE CASE THAT AN AGRICULTURAL BUILDING IS PART OF THE CURTILAGE OF A FARM HOUSE (ALTHOUGH THIS DOES NOT NECESSARILY MEAN THAT THE LAWFUL USE OF THE BARN IS ANCILLARY RESIDENTIAL)

BARN

GARDEN EXTENSION

GARDEN

IF THE BARN WAS IN SEPARATE OWNERSHIP AND FUNCTION FROM THE LISTED DWELLING AT THE TIME OF LISTING IT WILL NOT BE IN THE SAME CURTILAGE

HOUSE

FIELD

ORCHARD

NOTE:

THE OWNERSHIP OF THE CURTILAGE MAY BE DIFFERENT FROM THE LISTED BUILDING'S CURTILAGE (i.e. the barn could have been sold on, but at the time of listing it was in one ownership and was used in connection with the farmhouse)

THE GARDEN AND ACCESS TO THE LISTED FARMHOUSE ARE CLEARLY IN THE CURTILAGE IN THIS CASE, BUT IF THE DRIVEWAY AND GARDEN ARE LARGE, THEY MAY NOT BE

IF THE ORCHARD WAS USED AS A GARDEN SPACE AT THE TIME OF LISTING IT MAY BE IN THE CURTILAGE OF THE LISTED HOUSE, BUT IF IT WAS AGRICULTURAL IT IS NOT

HEALTH WARNING:

THERE IS NO DEFINITION OF CURTILAGE AND EACH CASE IS INDIVIDUALLY ASSESSED. FACTORS TO BE CONSIDERED INCLUDE:
- Physical layout
- Ownership
- Use or function

FIGURE 5.2 This sketch illustrates examples of the issues that need to be weighed up when assessing what is the curtilage of a listed building, in this case a farmhouse, surrounding land and adjoining barn.

unlisted, just because it is part of the historic curtilage of a listed building. If it is necessary to establish the curtilage of a listed building it is wise to consult the local planning authority early on in the commission.

Any curtilage buildings or objects constructed later than July 1 1948 are not protected by the listed designation (unless they are attached to the listed building or a curtilage listed building) and can be freely altered internally. However permission may be needed to alter them externally unless it is **permitted development** (see Chapter 7 for more details).

Conservation areas

Conservation areas are designated by local authorities. English Heritage can also designate conservation areas within London after consulting with the relevant borough council. In exceptional circumstances the Secretary of State can designate a conservation area, but this is normally done only where it is of national significance. The areas have to be of 'special architectural or historic interest, the character or appearance of which is desirable to preserve or enhance'[6]. The buildings inside a conservation area will not necessarily be listed, but some might be. The group value of the external appearance of the buildings and landscape is the main concern for the local authority, and control extends to trees and advertisements, as well as buildings.

There are now over 9,300 conservation areas throughout England, and over 650 in Scotland, 500 in Wales and 50 in Northern Ireland. Within these conservations areas there are in excess of a million buildings. The size of a conservation area can vary from quite small, with a few buildings included, to large ones containing over a thousand buildings. Every council in the country has at least one conservation area.

Within conservation areas, the local planning authority has powers to introduce additional local regulations called Article 4 Directions, which can be used to remove or reduce normal permitted development rights.

The council also has additional powers to ensure that buildings within the conservation area are maintained in order to preserve the character of the area. These include powers of compulsory purchase and control over demolition.

Development can still take place within conservation areas but the local planning authority has a statutory duty in regard to preserving or enhancing the character or appearance of the area. In addition to normal planning and listed building consents, consent is normally needed within the designated area for the following:

- Any works to a tree in the conservation area.
- Demolition of any building or structure (including boundary walls) in a conservation area including unlisted buildings, unless the building is less than 115m^3 and its loss will not affect the character of the area (if for instance it is hidden behind a garden wall out of public sight). The need to also get a conservation area consent as well as a planning consent was abolished in the Enterprise and Regulatory Reform Act in 2013.

The council may extend their controls, under Article 4 Directions (2) to reduce permitted development rights. These Article 4 Directions often include:

- Restricting the size of extensions which would normally be considered permitted development
- Restrictions to style and types of advertisements
- Restrictions on the erection of satellite dishes
- Restricting paint colours of external materials

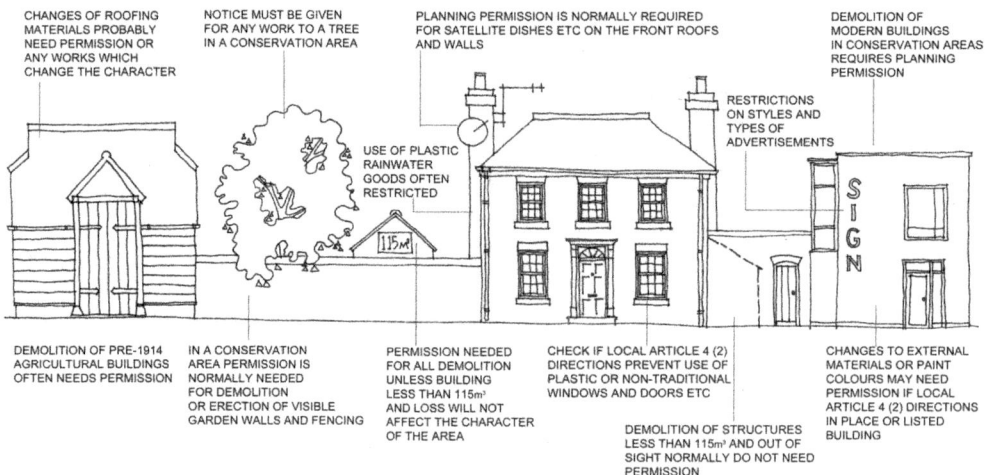

FIGURE 5.3 This sketch illustrates works that are likely to need planning permission in a conservation area.

- Preventing the use of plastic or non-traditional windows
- Demolition of some means of enclosure such as walls
- Demolition of pre-1914 agricultural buildings

Inside the conservation area planning permission is not be needed to carry out internal work on historic, but unlisted, buildings unless the changes could have an effect on the character of the conservation area.

When working in a conservation area it is sensible before advising clients, to become familiar with the local Article 4 Directions by contacting your local Planning Authority or consulting their web pages.

Planning permission will often be needed in the normal way for external alterations to any buildings or structures within the conservation area where they materially alter the building. This may even include changes to windows or external joinery where they affect the character of the area.

Buildings of local interest

Many local planning authorities have recognised that there are historic buildings which have no statutory protection, and yet their loss would be significant to the character of an area. These may be buildings outside a conservation area, but which are an asset to the community because of their appearance or historic significance. Councils faced with the problem of protecting these undesignated buildings often compile a register referred to as a 'local list', 'supplementary list', or 'list of buildings of local interest'. Historically, the buildings on local lists were sometimes referred to as Grade III listed buildings, although this grade is now obsolete

Some local planning authorities have gone on to use their powers under Article 4 Direction (2) to create their own local regulations which require notice to be given to the council prior to any demolition of 'Local List'' buildings. In these cases, if the council feels strongly enough that the building should not be demolished, they can again use the Article 4 Directions (2) to create a regulation to control its demolition. Therefore if it is proposed to demolish an older building which is of some significance to the character of an area, it should be first checked to determine whether it is on any local lists prior to proceeding.

Non-designated heritage asset

Some individual unlisted buildings are worthy of special consideration when changes which need a planning application are proposed. Their architectural and historic value falls below the standard required to make them a designated listed building, yet they are valued by their communities. Such buildings may not have any protection nor be included on any Local Lists or in a conservation area. Their value is sometimes only realised when a planning application is made and there is a strong objection from the local community to the proposals on the grounds of the building's unrecognised heritage value. Planners call these buildings non-designated heritage assets and are likely to treat any planning application affecting them with the same care as a planning application in a conservation area, or if on a local list. The National Planning Policy Framework ensures that the alteration or demolition of a non-designated heritage asset is a material consideration in any planning application. Often such buildings are spot-listed or added to a local list following the planning application to provide some protection for the future.

There is little one can do to establish if, through the planning process, a building may be deemed as a non-designated heritage asset, apart from consulting the planning officer for his or her opinion. However a thoughtful architect will have analysed the building's significance to the wider community prior to proposing changes and be able to assess the risks. If there is a significant risk that the building will be recognised as a non-designated heritage asset it is good practice to warn the client of these risks and the effect it may have on the outcome and length of the planning process.

Historic parks and gardens

English Heritage has compiled a 'Register of Historic Parks and Gardens Interest in England'[7]. Planning authorities have a duty to consult this register and English Heritage if there are proposals which would have an effect on any of these historic parks and gardens or their setting. They also have a duty to try and protect these areas. Plants and trees grow and die, and clearly gardens are not static objects. Historic parks and gardens therefore obviously cannot be guaranteed to be preserved without change.

There is no mechanism for obtaining consent to work on these parks and gardens, but architects need to be aware if their proposals are likely to affect any such land on the register, since this could be a valid reason for a planning consent being refused. Examples where the status of a designated historic park and garden may be relevant include where there is a planning application for a new golf course, or to allow an engineering operation such as a new road. The impact of these proposals on the surviving significant fabric of the historic landscape would become a material consideration for the planning application.

Historic parks and gardens often contain listed buildings, and this will ensure any structures within the garden curtilage of the property such as statues, walls, gazebos. Terraces also have protection and will need listed building consent to be altered. Trees may also have tree preservation orders, or be within churchyards or in a conservation area, and this provides some control on their management.

The criteria that English Heritage use to select suitable parks and gardens to be included in the register includes:

- That the park, or garden was created before 1750 and the original layout is still in evidence even if only a small part
- Sites where the main phase of development was between 1750 and 1820, and where enough of the original landscape survives to reflect the original design intent
- The best sites, parks and gardens where the main phase of development was between 1820 and 1880
- Parks and gardens laid out between 1880 and 1930 which are of high importance and survive largely intact. These are probably gardens of great aesthetic merit which were influential in the development of taste, culture or literary references
- Exceptionally important parks and gardens laid since the Second World War

- Parks and gardens which survive largely intact, and are representative examples of layout, or the work of the nationally important designer
- Sites which have an association with a significant person or historic event
- Sites where there is a strong group value as part of the wider landscape which may include listed buildings, or be an example of an important town planning scheme

The register is also now being used to grade cemeteries which are of national importance.

The same grading system is used for Historic Parks and Gardens as is used on listed buildings; parks and gardens of exceptional interest are Grade I, those of great interest are Grade II*, and those of special interest are Grade II.

Battlefields

Since 1995 there has been a register of some 43 English historic battlefields prepared by English Heritage, and there are proposals to create a similar register for Wales. It is used by local authorities in a similar way to the register of Historic Parks and Gardens, and is material consideration when planning applications are submitted on historic battlefield sites on the register[8].

World Heritage sites

In 1984 Great Britain signed up to the World Heritage Convention which was promoted by UNESCO. This convention seeks to protect cultural and historical sites which are of international significance. There are currently 28 sites in Britain, and the list is growing, and it includes whole areas containing listed and unlisted buildings. They include the old and new towns of Edinburgh and Bath, the monastic areas of Canterbury, maritime mercantile Liverpool, maritime Greenwich, single buildings like Blenheim Palace, as well as building complexes such as the Derwent Valley Mills where the modern factory system was born.

In practice, designation of a building or area as a World Heritage Site does not lead to any additional need for special applications to allow building works. All these sites are protected by the standard protection afforded by conservation area status, scheduled monument or listed building designation. The local authorities therefore use these normal development management measures to ensure the World Heritage Site is preserved and its character is not affected by inappropriate development.

Notes

1   DCMS and the Welsh Assembly Government (2007), *Heritage Protection for the 21st Century*, available at www.gov.uk
2   HM Government (1913), Ancient Monuments Consolidation and Amendment Act, s.22
3   HM Government, (1979) Ancient Monuments and Archaeological Areas Act 1979, available from www.legislation.gov.uk
4   HM Goverment (1983), National Heritage Act and for Scotland the National Heritage (Scotland) Act (1985), available from www.legislation.gov.uk
5   HM Goverment (1990), Planning (Listed Buildings and Conservation Areas) Act 1990, available at www.legislation.gov.uk

6   HM Goverment, Planning (listed Building and Conservation Areas) Act 1990, Section 69 paragraph 3
7   The register can be accessed from, The National Heritage List for England, available at www.english-heritage.org.uk
8   The registered Battlefields are listed on English Heritages web site, www.english-heritage.org.uk

Further reading

> Department of Culture, Media & Sport (2013), *Scheduled Monuments Policy Statement – Identifying, protecting, conserving and investigating nationally important archaeological sites under the Ancient Monuments Act 1979*, London, DCMS, also available from www.gov.uk
> Department of Culture, Media & Sport (2010), *Principles of Selection for Listed Buildings*, London, DCMS, also available from www.gov.uk
> English Heritage (2010), *The Register of Parks and Gardens*, available at english-heritage.org.uk
> English Heritage (2011), *Designation Listing Selection Guides series*, London, English Heritage, available from www.english-heritage.org.uk/publications. These guides include: Agricultural Buildings, Commemorative Structures, Commerce and Exchange Buildings, Culture and Entertainment, Domestic 1 Vernacular Houses, Domestic 2 Town Houses, Domestic 3 Suburban and Country Houses, Domestic 4 Modern House and Housing, Education Buildings, Gardens and Park Structures, Health and Welfare Buildings, Industrial Structures, Law and Government Buildings, Marine and Navel Buildings, Places of Worship, Sports and recreation Buildings, Street Furniture, Transport Buildings, Utilities and Communication Structures
> English Heritage (2011), *Understanding Place: Conservation Area Designation, Appraisal and Management*, London, English Heritage, available from www.english-heritage.org.uk/publications
> Harwood, R. (2012), *Historic Environment Law: Planning, Listed Buildings, Monuments, Conservation Areas and Objects*, Builth Wells, Institute of Art and Law (not yet revised following NPPF)
> HM Government (1979), *Ancient Monuments and Archaeological Act 1979*, available from www.legislation.gov.uk
> HM Government (1979), *Planning (Listed Buildings and Conservation Areas) Act 1990*, available from www.legislation.gov.uk
> Menuge, A. (2010), *Understanding Place – Historic Area Assessments: Principles and Practice*; London, English Heritage, available from www.english-heritage.org.uk/publications. This book is still considered by English Heritage as Planning Guidance although it has not been updated to take account of the new NPPF
> Mynors, C. (2006), *Listed Buildings, Conservation Areas and Monuments (Fourth Edition)*, London, Sweet & Maxwell. (This book is currently a principle reference work for planning consultants, lawyers and barristers who are involved with legal issues relating to heritage assets, and recommended reading for specialist conservation architects. However the fourth edition does not take account of recent legislative changes including the NPPF, and the Enterprise and Regulatory Reform Act)
> *Heritage Protection for the 21st Century*.
> English Heritage (2011), *Understanding Place: Conservation Area Designation, Appraisal and Management* (some references in this document are out of date since the NPPF but it is still the best reference available currently).

CHAPTER 6
ASSESSING A BUILDING'S SIGNIFICANCE

As explained in Chapter 3, architects and surveyors have a professional duty to study and analyse the significance of any older buildings they are proposing to change, no matter if they have a designation or not. Only after 'reading' and studying a building, and understanding if it has any features which are of aesthetic, historic, cultural or spiritual value for the past, present and future generations, can we balance the need for change against the significance of the existing fabric. It is also our professional duty and good practice to convey our analysis of the significance of the fabric to planners and conservation officers and councils, so they can evaluate our designs with a full understanding of the building's value, and weigh up the need against harm.

Some buildings gain significance because of small architectural features, for example a rare surviving medieval wall painting or a fine fireplace. Others are significant as a complete example of the work of a particular famous architect or are rare surviving building types from a previous age. Others gain significance by association with historic events and people.

The significance of some other buildings becomes more marked with time, as towns and cities evolve and buildings are converted, demolished or altered. This leaves fewer examples of once plentiful building types. So what is currently insignificant may become significant as the building form becomes less common. In our northern cities industrial mill buildings were once so plentiful that they were not considered worth listing, but now, after wholesale clearances in some cities, those which survive may be worth listing as significant historic structures speaking of another age.

Redundant barns and other farm buildings are often converted to provide dwellings. However, there are now so few left unaltered that the historic significance of the remaining barns has risen to such a point that the heritage authorities are reluctant to allow more residential conversions. This can be frustrating to a client who looks around a neighbourhood and sees many conversions of agricultural buildings carried out a decade or so before, but is now not allowed to convert his building to a house.

Many owners of listed buildings are oblivious to the significance of their property; they may live in a 400-year-old timber-framed house and, for example, resent not being allowed to remove walls in order to enlarge a room. It is often necessary to guide clients to an understanding of the significance of the features, and therefore the value of their homes, and to not waste time proposing alterations which damage important fabric and which will most likely be rejected by the planning authority. It is important to ensure that owners understand the special character and significance that their property holds, or they will not be able to comprehend the logic of any proposed scheme.

Researching the significance of a historic building

The first step for any architect working on a historic building should be to establish the significance of the building: architecturally, culturally, socially, historically and aesthetically. This assessment will inform the design process by clarifying how much change a building can absorb without damaging the things that make it significant. One of the first tasks therefore for an architect commissioned to work on a historic building is to compile information for a Heritage Statement.

Proportionality of evidence

Current planning legislation is clear that the level of evidence needed regarding the significance of a building that should accompany an application to alter a heritage asset will need to be proportionate to its significance, and should be no more than is sufficient to understand the potential impact of the proposals on its significance. Therefore the time spent assessing the building will vary, and should be proportional to the significance of the asset being considered and the extent of the alterations. For example, even minor changes to a Grade I listed building is likely to require a fuller assessment than a proposal for alterations or additions to a Grade II house.

For some domestic Grade II listed buildings there may be less documented historical information available, and the history study and significance report will depend on the architect's ability to 'read' the building, and present an authoritative judgement of its qualities. Not having good written evidence is not a reason for taking less care, and a well-perceived and knowledgeable study of the building itself may well become the updated listing description for future reference.

An application which is largely about repairing decayed fabric and involving little or no change to the character of a building may only need a simple Heritage Statement since the impact on significant fabric would be very limited. However even for this type of repair work the architect will need to establish as a minimum the age, the significance and materials needing repair.

For a major re-ordering however, for example to Grade I church, it can be reasonable for the authorities to require a very detailed assessment of the building's history and significant features. The Statement of Significance (or Heritage Statement for non-ecclesiastical buildings), containing this information, may need to be a lengthy document to convey a full understanding of the property. The church will have social significance to the local community which has changed over time. It may have had a complicated evolution with extensive alterations and extensions, it may contain artistically important memorials, furniture or stained glass, significant archaeology may be hidden beneath the floors. The **evidence value** of the building will be large and all this information will need to be understood and conveyed to all the parties involved in assessing whether the building can absorb the changes proposed without damaging the significance of the whole.

Sources for the evaluation of significance

For most projects the architect can evaluate the significance of the building him/herself and dovetail the many sources available with their own observations on site to develop a good understanding of the property. From this research the architect can write the heritage statement (or statement of need for a church).

These sources may include:

- **Listed Building Grades**

  The listed building Grade be it I, II* or II is clearly a strong indication of a buildings overall significance and are a strong clue to the amount of research which will be needed on a project.

- **The National Heritage List for England**

  One of the first things to check is whether the building is listed, then read the list description. The listing details can be found in England on the National Heritage List for England via the English Heritage website (www.list.english-heritage.org.uk), or can be obtained at the local library; they can also be obtained directly from the local planning authority. When searching for a particular building on the English Heritage website it is best to search by the location of the building and then check the list that the search brings up. The database is fussy, and if the name of the building is not entered exactly as noted in the data it will reject the search.

  The listing will have a brief description of the significant features noted at the time of listing. Sometimes it includes sources about the building in the references at the bottom of the description and these can be followed up. It may even state significant features which have made the building worthy of a designation. It is widely accepted that the current listing descriptions are often a poor guide. Few buildings were inspected internally, and little documentation was consulted in the hasty re-listing in the early 1980's following the Firestone Factory scandal[1]. Therefore the architect should read them with caution and fully examine the building to learn a lot more.

- **Heritage Gateway**

  In England the architect can also consult the Heritage Gateway website (www.heritagegateway.org.uk), which is a good starting point to find other sources about designated heritage. From here it is also possible to also find the National Heritage List, which contains the listings of individual buildings.

  The Gateway can be used to check for Historic Environmental Records (HER) which are normally held by county councils, district councils and unitary authorities. These records include local records such as archaeological finds, historical aerial photographs and records kept about listed buildings. Some counties have more sources listed than others, and the Heritage Gateway site gives the location of the office where the HER is kept, and how to get in contact. However, there will be some variation in the number of sources listed depending on the county/authority, and there may be a charge to de-archive records.

  Another source that the Heritage Gateway searches is the National Monument Record (NMR), which contains a source of public records of over 12 million photos, plans, drawings, reports and publications on archaeology, listed buildings, aerial photography and social history.

  The Heritage Gateway also checks the English Heritage Images of England site www.imagesofengland.org.uk, which has recent photographs of most listed buildings in England alongside a copy of their list description.

- **Owner's records**

  The architect should interview the owners of the building and find out what they know of its history and ask if they have any sources that can be followed up. They may have collected old plans, prints and photographs which can be used to give clues to a building's development, especially if they are dated.

- **Site inspection**

  The architect should do a thorough site inspection and survey, and whilst surveying the building, consider how it has grown and changed. Logical deductions can be made from the observations, developing a picture in the mind of how the building has evolved. Which parts look like extensions and what looks original? Clues can often be found from careful examination, for example looking to establish if walls are built on top of each other etc.

  It may be necessary, to make small holes in a structure to establish the age of it. However if the building is listed, consent may be needed to do this, so it is best to ask the Conservation Officer if they are happy for the structure to be opened up.

- **Measured survey**

  The architect should prepare a measured survey of the building which can be used to explain its development. It can be very time consuming to survey a historic structure which may not be straight or regular but there are now modern methods of measuring which can be very useful when analysing a listed building, including laser cloud surveys, and rectified photographs (The RIBA Good Practice Guide: Building Condition Surveys is a good guide on this subject)[2]. The product of these survey techniques when imported into a CAD system provide a very accurate, scalable and measurable document, over which notes can be made, or drawings developed.

- **Old maps**

  The architect should look at old Ordnance Survey maps which are easily bought on-line from Ordnance Survey and other suppliers. The Ordnance Survey has been surveying Britain since the early nineteenth century, and the maps were regularly updated as urbanisation was taking place. Comparing these maps at spaced time intervals of 25–50 years can give important clues to the age and development of the building complexes. They can establish when a structure was constructed or demolished and give clues to the historic curtilage.

- **Conservation officers**

  It is worth asking the conservation officer if they have any records on the property or can direct to any local sources. If listed building or planning applications have been made on the property in the past, the conservation officer may have a considerable understanding of the building's history.

- **Old planning or listed building applications**

  Most Local Authorities now have web-based search engines where the past planning history of a property can be tracked and, if fortunate, old application documents may be viewed online. Sometimes these can also be accessed via the Planning Portal at www.planningportal.gov.uk.

- **Local sources**

  Check local sources like the library, county and metropolitan authority archives, parish records, local history society and books with historic photos, old post cards, and local histories for any records.

- **Diocesan records**

  The Diocese of the Church of England and their registries keep archived records of the faculties granted to change church buildings, and these records can be very helpful if the architect is altering a listed Anglican Church. These records often contain plans, drawings and specifications and are normally available from the nineteenth century onwards. Other denominations have similar records of applications and consents.

- **The National Archives**

  There are records held in the National Archives in Swindon and Kew which have been collected by many public bodies. They include the old photographic archives and the Records of the Royal Commission on Historical Monuments (England) which gathered information on historic monuments and buildings in the period from 1908 until they were absorbed into English Heritage in 1991. The National Archives have an index to their collections that can be searched online (www.nationalarchives.gov.uk/about) and copies of archives can often be reproduced and sent out to researchers, or book an appointment go in person to consult them.

- **Magic**

  Magic is a web-based interactive map which brings together environmental information from English government agencies responsible for rural policy-making including Defra, DCLG, English Heritage, Natural England, Environmental Agency, Forestry Commission. The information is displayed on a map with links to other sources. It can be used to identify Historic Parks and Gardens, Scheduled Monuments, Conservation Areas and Landscape Character Areas etc. www.magic.defra.gov.uk

- **Reference books**

  There are also standard reference books such as The Pevsner Architectural Guides (originally called the Buildings of England Guides) give information on most significant historic buildings of importance, from cathedrals to industrial architecture. These guide books to the architecture of the counties of the British Isles were begun in the 1940's by Sir Nicholas Pevsner but now often have other authors or co-authors[3].

- **Biographies of architects**

  If the building is Victorian and the architect of the building known, there is a very guide published by RIBA, *The Directory of British Architects, 1834–1914*[4], which lists the known details of most architects operating in Britain during this period. It is an expensive book but to a church architect dealing with many Victorian churches it can be a very useful reference document.

- **Church Plans on Line**

  If working on historic Church of England churches there is a very useful database of historic plans and elevations available free on the web at the Church Plans on Line website at www.churchplansonline.org. It shows at a small scale some 13,000 digitised plans and drawings which are part of the Lambeth Palace collection. Larger scale drawings can be ordered for a modest fee.

- **Consulting accredited conservation architects**

  If the architect does not feel competent to analyse a building's significance when making alterations to a complex listed building, it may be worth employing a specialist. The assessment of a building's significance is one of the key skills required of an Accredited Conservation Architect and many RIBA Accredited Conservation Architects regularly act as a consultant to other architects in the preparation of heritage statements, **conservation management plans** etc. They can be very helpful to less specialised architects by using their skills as a **Competent Person** of evaluating the significance of a building, and can advise in the preparation designs and negotiations with conservation officers. The use of experienced Accredited Conservation Architects who bring their specialist knowledge, and can be objective in their assessment of the significance of the fabric, is often welcomed by the authorities.

- **Other consultants**
  The architect can also employ an independent architectural historian to research the history and significance of the building; there are now a number of specialist firms who will undertake this work. There are also house historians who can be found online, who specialise in researching the history of domestic houses. Historians can be helpful, but their research is often focussed on the ownership of a property by tracing details found in property deeds and wills, whereas specialist accredited conservation architects often have a greater understanding of the needs of a fellow architect when assessing a building's significance.

There are also many archaeologists who are specialists in recording historical buildings and assessing the historical fabric. They often work in a particular region and develop extensive knowledge of local historical vernacular building technologies. They will do documentary research, and know their local archives very well, but their special skills are in directly examining the built fabric. They can be especially helpful in making judgements about significance in early and complex historic buildings which have developed over a long period. As for experienced accredited conservation architects, they can help deduce the building's development, and relevant dates, provide art-historical dating from the style and design of features, provide structural understanding of construction from different periods and put the building into context with other similar buildings from the region.

In early discussions with the conservation officer one may even be asked to commission archaeologists to carry out a desk based archaeological assessment so there is a clear and authoritative understanding of the extent of the historic fabric and the risks faced by granting consent. For example a major alteration to a medieval church will almost certainly need an archaeological assessment. The use of professional archaeologists can be expensive but as with using accredited conservation architects their judgements and advice will be seen as a reliable source in any planning process making documentation more authoritative.

Conveying research within the heritage statement or assessment

Once a clear understanding of the significance of a historic building has been established from research this information needs to be conveyed to the client and the heritage authorities.

If the architect is making a listed building application this information is normally conveyed in a heritage statement (sometimes referred to as a historic statement or heritage impact assessment, or 'heritage assessment', and when dealing with a church as a statement of significance). It may also be necessary to produce a heritage statement for any older building or non-designated heritage asset. The National Planning Policy Guidance suggests that the heritage assessment form part of the design and access statement. If the architect is applying for consent on a church covered by an ecclesiastical exemption from the normal listed building application process, the equivalent document to the secular heritage statement is called a statement of significance.

A conservation officer is more likely to look on proposals favourably and be supportive of it if a good understanding of the history and significance of the building has been clearly demonstrated. Indeed the conservation officer is unlikely to be an expert on the building and will often tend to assume that most sections of the building are of historical and architectural significance unless demonstrated otherwise. The burden of proof is with the applicant. This can only be done when there is a clear understanding of the significance of the fabric affected. For example, the removal of a wall in a medieval timber framed building may be desired to make a larger more useable space. The wall could be contemporary to the original building with rough hand cut stud timber, and if so, the conservation officer is unlikely to agree to the proposal. However, it could be made from straight mechanical sawn timber studs and covered with plasterboard and therefore it is obviously more modern, and of little significance. Consent therefore could be expected to be given for its removal and it would be necessary to convey the evidence of the construction of the wall, with a photograph of the studs, in the heritage statement.

Heritage statements are normally divided into sections which cover a general description of the building, discussion about its significance relative to buildings of a similar typology and finally a detailed assessment of the alterations and their effect on significant fabric, with an appendix detailing reference sources. Within the individual sections it is good practice to include the following information:

1. **General description of the building and its significance**
- A general introduction giving the current use and listing grade of the building
- A description of the building's setting
- An introductory brief schedule or list of the specific changes proposed, so the reader has this in mind when considering the significance of the fabric
- A description of the history of the building and the history of the immediate locality so the proposals can be read in context. This section may include:
  — When was the building first built and how has it changed over time?
  — Is it known who the architects, builders, artists, and craftsmen were who were involved with the building?
  — How have the plan, building form and spatial quality developed over time?
  — What are the construction techniques and materials used in its construction and when were they used?
  — What is the history of any buildings in the curtilage of the building and the immediate area beyond?
- The history needs to also cover the social history of the building giving:
  — Present use of the building and curtilage as well as its historic use and how this contributes to the buildings wider social significance.
  — How has the community around the building has changed over time?
  — Whether there are any significant events or personalities associated with the building.
  — Whether there are important communal memories associated with the building.
- Where an application will have an effect on the external appearance of a building there should be a general description of the setting of the building including:

- Identifying whether there are any distant or close views which are valued by the local community; if there are photographs of the building from these vantage points they should be included.
- Is there anything known of the history of the landscape and surrounding buildings in its curtilage?
- Are the adjacent buildings similar, complementary or contrasting in style, age and construction?
- How are the boundaries around the building formed?
- Are there any monuments surrounding the building?
- Are there any known archaeological remains or ancient monuments in or adjacent to the building?
- Description of any significant features of the building, especially if they are mentioned in the listed building description.
- Descriptions of the building's fixed contents and their significance, especially when they are historic or contemporary with the building, or mentioned in the listing description.
- A general description of the building's condition and any risks to the preservation of its significance.
- Specific descriptions of any parts of the fabric which will be affected by the proposals, such as any information known about their age, construction, manufacturer and design.

2. Analysis of the relative significance of the building

It is good practice in the heritage statement to evaluate the impact of the proposed changes in relation to their significance and the value of the fabric. In assessing significance the following graduations may be used, as suggested in the British Standard 7913 (Guide to conservation of heritage buildings)[5] to assist in assessing likely capacity for change:

Heritage value:
- Very high – important at national to international levels
- High – important at regional or sometimes higher
- Medium – usually of local value but of regional significance for group or other value (e.g. vernacular architecture)
- Low – of local value
- Negligible – adds little or nothing to the value of a site or detracts from it

3. Analysis of how the Alterations Affect Significance – The Impact Assessment

Following the general description of the building and its significance, it is good practice to 'focus' on the specific parts of the building and curtilage which will be affected by the proposals; this is often referred to as the **impact assessment**. In this last part of the statement it is usual to include the following:

- Identify the parts of the building and surrounds that will be affected by the proposals, analysing their significance and why the changes are necessary.
- Identify if the impact and level of harm of the proposals on the significant parts or whole building is low, moderate or substantial. The building may be a significant listed building but if alterations can be restricted to parts of the building which are modern or of little architectural/historical significance, the impact of the proposals will be low consent is much more likely to be granted.

SIGNIFICANCE					
VERY HIGH	NEUTRAL	LOW	MODERATE/ HIGH	HIGH/VERY HIGH	VERY HIGH
HIGH	NEUTRAL	LOW	LOW/ MODERATE	MODERATE/ HIGH	HIGH/VERY HIGH
MEDIUM	NEUTRAL	NEUTRAL/ SLIGHT	LOW	MODERATE	MODERATE/ HIGH
LOW	NEUTRAL	NEUTRAL/ SLIGHT	NEUTRAL/ LOW	SLIGHT	LOW/ MODERATE
NEGLIGIBLE	NEUTRAL	NEUTRAL	NEUTRAL/ LOW	NEUTRAL/ LOW	LOW
	NO CHANGE	NEGLIGIBLE	MINOR	MODERATE	MAJOR

IMPACT OF CHANGE

FIGURE 6.1 The magnitude of the impact of a change can be assessed by plotting the value of the heritage and impact of the change as illustrated in the British Standard BS 7913.

The British Standard 7913 (Guide to conservation of heritage buildings) suggests that an assessment is made on the impact of any changes against the heritage value of the building using the following scale:

Impact:
- No change
- Negligible
- Minor
- Moderate
- Major

A scale such as this will need to reflect the significance of the building, as what could be seen as a minor alteration in a Grade II listed building such as the creation of a single new opening may be a moderate or even major change in a Grade I listed building.

Where possible, the architect should explain how the impact of the proposed works on the character of the building can be mitigated, for example by re-locating disturbed fabric such as an old door. It is also usual to explain how reversible the changes proposed are. It can also be helpful to show that several options have been examined, looking to minimise the impact of proposals. After all, if historic fabric is demolished it is lost forever, as it can never be **authentically** replaced.

4. Sources and Appendix
Finally it is good practice to list the sources used to prepare the statement, and it is furthermore useful to include a copy of the listed building description. It can also be helpful to give short biographies of the architects, artists or craftsmen who have worked on the building, if they are known. This will help you and the planners assess the significance of the work and help judge its importance. Establishing that a well-known architect worked on

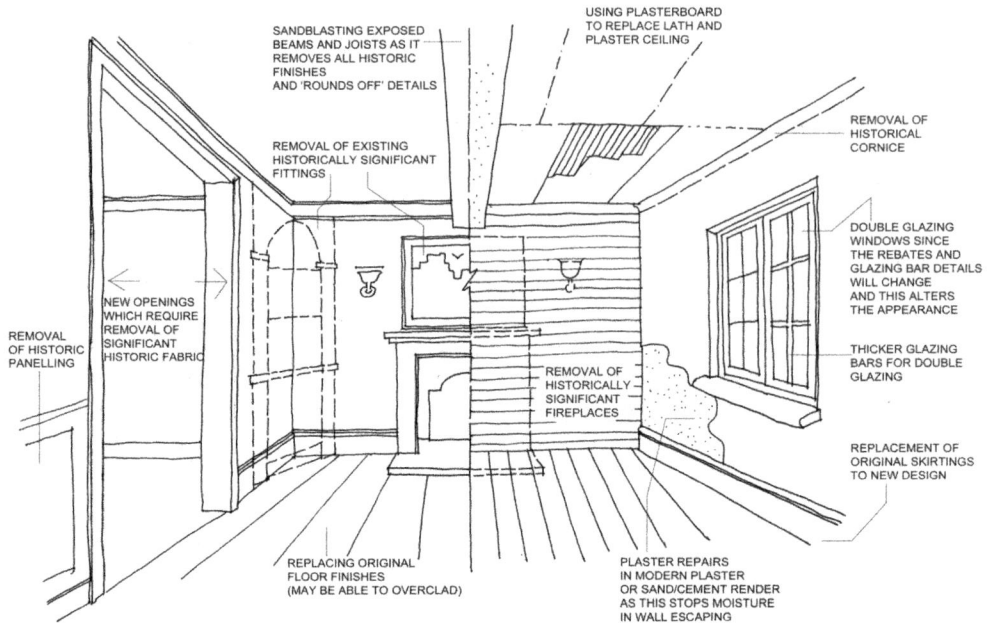

FIGURE 6.2 This sketch shows items of work which are commonly included in applications for listed building consent but are likely to be refused because the changes will impact on the significance of the building.

the building will normally be seen to increase its significance. Of course many, if not most, historic buildings are constructed by unknown designers or were part of a local vernacular.

The heritage authorities will be weighing up the significance of the fabric to be altered against the need. Their remit is to ensure the preservation of significant fabric, but it is also to manage and encourage change so that it is sympathetic and ensures the building's upkeep. They will have expertise in evaluating the significance of old buildings, and where a Conservation Officer is unsure they will often call in expert advice from English Heritage, Historic Scotland etc or one of the National Amenity Societies such as SPAB, the Victorian Society etc. who have panel of experts. The foundations of a design will therefore need to be grounded in a firm understanding of the building's history and its significance.

The heritage statement is also sometimes expanded to include specialist advice from other consultants etc and this is explained in Chapter 7.

Conservation planning

Conservation planning is a process which is used to establish the most suitable means to sustain significance where changes are needed or proposed. For example a conservation planning approach may be used where buildings, or a group of buildings, are identified as being 'at risk' of loss or decay and their owners, or the local planning authority, or special partnerships of interested bodies are formed to bring them back into use. The use of conservation planning techniques is not new, but the production of a written conservation statement, plans and management plans has developed since the 1990s and has been promoted by

English Heritage and the Heritage Lottery Fund. The process has four logical steps:

- Understand the historic building: its origins, historic use, setting, construction, development, current use and movement.
- Assess and articulate its significance: what is important about the building and to whom.
- Articulate what are the issues and threats to the buildings significance: change, use, wear and tear, decay etc.
- Establish appropriate conservation management policies to address the issues and threats: for its use, maintenance, repair, whilst always minimising harm.

Conservation statements

The initial step in conservation planning is often the production of a conservation statement or a strategic plan which provides an overview of the building or group of buildings, history, heritage value and a summary of its significance and issues facing its future management. They use readily available information and may identify the need for more research or specialist studies. They may be sufficient to guide decision-making on relatively straightforward cases, but more important cases may need the more careful analysis of a conservation plan. They can also be used as an initial assessment on more complex cases and be used to inform the commissioning of more detailed conservation plans.

Conservation plans

Complex projects may require detailed conservation plans to resolve complex issues and where conflicts exist between the management of the building and preservation of the fabric. A conservation plan will describe in detail the fabric, its historical evolution, setting, aesthetic merit and evaluate in detail the relative significance of individual elements of the fabric. The plan will then assess the capacity of the fabric to withstand change, including planned maintenance, necessary repair, or changes of use and development. It is likely to draw on research as well as a careful analysis of the physical fabric and is normally done by experienced conservation accredited architects.

Conservation management plans

Conservation management plans provide more detailed guidance than conservation plans and are normally produced by accredited conservation architects who have the suitable expertise. They probably will contain specific guidance about the management and maintenance of the historic building to ensure its significance is maintained. Large grant making bodies such as English Heritage when making grants often require that a management plan is produced and the actions detailed in it carried out.

Notes

1   The Twentieth Century Society, *A Brief History*, available on their web site on, c20society.org.uk. Retrieved 7 July 2014
2   Hoxley, M. (2009) *Good Practice Guide: Building Condition Surveys*, RIBA Publications

3   Buildings of England, Scotland, Wales and Ireland Series, Yale University Press
4   Brodie, A. (ed.) (2001), vols. 1 & 2 of *Directory of British Architects 1834–1914*,
     Bloomsbury Academic (vol. 1), Continuum International Publishing Group (vol. 2)
5   British Standards Institution (2013), *BS 7913 Guide to the Conservation of Historic
     Buildings*, London, BSI

Further reading

>   British Standards Institution (2013), *BS 7913 Guide to the Conservation of Historic Buildings*,
    London, BSI. (Authoritative source for terminology, the assessment of significance and the
    production of conservation and conservation management plans etc)
>   Church Care (2012), *Guidance Note Conservation Management Plans*, London, Church
    Buildings Council, available from www. churchcare.co.uk
>   Clarke, K. (ed.) (1999), *Conservation Plans in Action: proceedings of the Oxford Conference*,
    London, English Heritage
>   Clarke, K. (2001), *Informed Conservation*, London, English Heritage
>   English Heritage (2013), *Conservation Basics, Practical Building Conservation Series*, Farnham,
    Ashgate Publishing Ltd
>   English Heritage (2008), *Conservation Principles, Policies and Guidance for the Sustainable
    Management of the Historic Environment*, London, English Heritage, available from
    www. english-heritage.org.uk/publications
>   Heritage Lottery Fund (2012), *Conservation Plan Guidance*, London, Heritage Lottery Fund
>   Insall, D. (2008), *Living Buildings; Architectural Conservation, Philosophy, Principles and Practice*,
    Musgrave, The Image Publishing Group
>   Prince's Regeneration Trust (2009), *How to: Write Conservation Reports*, London, Prices
    Regeneration Trust, available from www.princes-regeneration.org
>   www.magic.defra.gov.uk is a web-based interactive map
>   www.churchplansonline.org
>   Individual County Guides, in series of books of, *Pevsner Guides to the Buildings of England,
    Scotland, Wales and Ireland Series*, Yale University Press

CHAPTER 7
APPLYING FOR CONSENT

Once the architect has researched the significance of the building and prepared the design proposal they will need to make an application that is appropriate to the type of historic structure. Normally on listed buildings the architect will need to apply for listed building consent for most works, but they may also need to obtain a planning permission as well. If in doubt they should consult the local planning authority or seek advice from a planning consultant.

Virtually all work to scheduled monuments will need scheduled monument consent. Some buildings are both scheduled ancient monuments and are listed, in which cases normally only the schedule monument consent will be needed to carry out works. Planning permission for any changes of use and extensions to the monument will also be needed in the usual way.

There is no special consent regime for historic gardens or battlefields, but when a planning application is made the local planning authority has a duty to inform and take advice from English Heritage before determining the application. The local planning authority have a primary statutory duty in regard to the significance of any heritage asset when determining any application, and English Heritage have the role to advise planning authorities over the significance of historic gardens and battlefields.

The need for listed building consent

Listed building consent will be required to carry out any works for the alteration, extension or demolition of a listed building in any manner 'which would affect its character as a building of special architectural or historic interest'[1]. This is an all-embracing requirement which covers more forms of work than are normally covered by a planning consent. In addition listed building consent is needed for work to the interior fabric and fixtures of the building, whether they are Grade I, II or II*.

If it is possible to demonstrate that only routine maintenance is being undertaken that will not affect the character of the building, this will not need consent. The law however is open to different interpretation by individual conservation officers, and it is worth consulting the local council before any work is undertaken without consent. On listed structures it is a criminal offence to undertake work which affects the character without consent. Many Local Planning Authorities employ enforcement officers who pursue offenders. Neighbours, local amenity societies etc are often sources of complaints to council when they suspect an offence has occurred. If a property has been altered in the past, records and photographs will have been kept on the council's file for future record. Solicitors now ask to see copies of consents when buildings are conveyed, and not being able to prove consent can affect values negatively, as the new owner will be taking on the risk of rectifying the non-conforming work, even though they were not responsible for the offence itself. Unlike breaches of the planning permission regime, there are no time limits for a local authority to take enforcement action to remedy any breach of listed building control.

It can be difficult to decide whether consent is needed for works which include repairs, painting, cleaning, advertisements, and works to buildings in the grounds of a listed building, but it is worth consulting the council if the architect is in doubt.

Repairs

Normal small-scale maintenance repairs to listed buildings which use materials virtually identical to those used originally should not need consent, since the character of the building will be maintained. Care needs to be taken choosing the materials and methods used for repairs as they can have a marked effect on a building's character. Often the original materials cannot be matched exactly, the original stone or brick or tile sources may have closed and substitute materials have to be used. This would normally constitute works affecting the character of the building and therefore need consent.

The loss of historic timber when repairing or replacing a window could also be seen as significant, and repainting a listed building in another colour can dramatically change the character of the building. Conservation officers will use their judgement to consider these grey areas, and it is therefore advisable to always consult them before deciding that a repair does not need consent.

Cleaning

Significant cleaning of listed building fabric nearly always needs consent, as the cleaning process will change the appearance of the building even if it is only restoring it to its former appearance. Cleaning processes such as washing stone or brick with water jets or solvents can also affect the historic fabric, for example there may be hidden iron cramps which rust because they get unusually wet in the cleaning process and this leads to them cracking the stone or brick coverings. Some smaller cleaning operations such as washing down windows, painted balustrades etc. are unlikely to affect the character or fabric and therefore should not need consent.

Sand blasting historic fabric, be it timber beams inside a listed building, brickwork or cast iron columns outside is a very aggressive form of cleaning and will remove all the historic layers of finishes which may be significant, and therefore nearly always needs consent.

Painting

Painting a listed building with a new coat of a similar paint in the same or very similar colour where it has already been painted would not normally be considered an alteration needing a consent in most of Britain. However in Scotland the painting of a listed building is often considered to be an alteration needing consent.

Changing the colour or using a different type of paint such as a modern synthetic paint on a wall which has previously been painted in a traditional material like distemper may need consent. There is more flexibility about the choice of interior paint colours, and many internal colour changes would be considered to be an alteration. The architect needs to be careful not to affect the character of the building adversely when dealing with significant fabric such as panelling or plaster mouldings where the colour of these features is part of a historic colour scheme which is of significance.

Painting inside or outside on hitherto undecorated materials, like undecorated render, stone or brick, does however need consent since this will alter the character of the listed building.

Satellite dishes and security cameras

In contrast to unlisted houses, the installation to listed buildings of satellite dishes, security cameras etc. normally needs listed building consent since in most cases their addition to a listed building will affect the character of the listed building.

Advertisements

The installation of a sign or advertisement can detract and alter the appearance of a historic building and therefore nearly always need listed building consent even when they would normally be permitted development on an unlisted building. The replacement of a sign on a listed building with a similar sign of similar design arguably will not affect the building's character and therefore probably does not need consent.

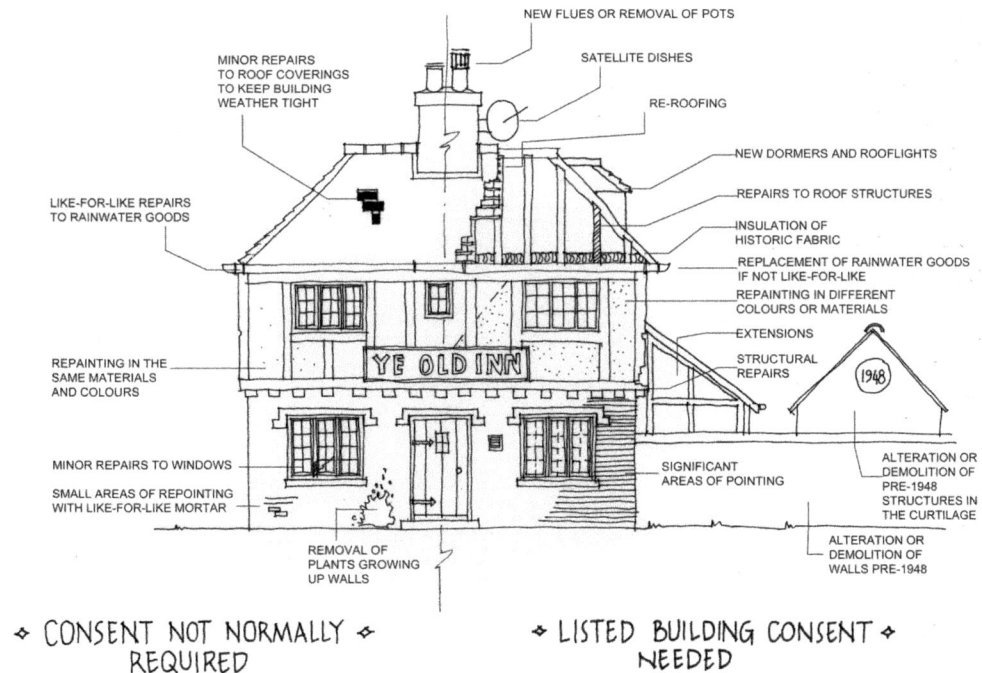

FIGURE 7.1 This sketch shows the types of work likely to need listed building consent

Alterations to buildings in the curtilage of a listed building

Planning law distinguishes between buildings in the curtilage of a listed building and erected before the first planning acts in 1948, and those built after 1948. Buildings in the curtilage constructed before 1948 may need planning permission and will need listed building consent to alter them externally. They also need listed building consent to alter them internally or to demolish them, if these changes effect the 'special character of the listed building', which is normally the case.

Buildings in the curtilage of a listed building which date from after 1948 can be altered internally without consent but may need a planning permission to alter them externally even when this form of alteration would have been permitted if the building were unlisted. They also normally can be demolished and altered externally without a Listed Building Consent, unless in the opinion of the council these works would "affect the special character of the listed building". This may be because they are very close to the listed building or look as if they were original (possibly a rebuilding after a fire). It is therefore worth checking the conservation officer's opinion before submitting an application.

Following the Enterprise and Regulatory Reform Act 2013, measures have been put in place to amend the Planning (Listed Buildings and Conservation Areas) Act 1990, which will allow applications to be made to exclude from listed building protection structures or objects attached to, or within the curtilage of the listed building where they can be identified as lacking special architectural or historic interest. It is envisaged that owners of listed buildings will apply for specific parts or features of the buildings to be excluded from listed building protection to reduce the need to make applications where there is no significant fabric present, for example the interior of a modern extension to a listed building which is of no historic or architectural significance.

Urgent works

Sometimes urgent work is needed for reasons of health and safety. The building may be about to collapse, an accident or fire may have damaged it. In these circumstances where urgent action is needed, it may not be possible to get consents in place before work is carried out. The council should be contacted as soon as possible and written notice of the actions to be taken given to the officers, and a dialogue established with the council as to the best way forward. It is important to note when urgent works are needed they are not exempt from normal planning and listed building controls. To proceed without written confirmation from the planning authority puts owners in great risk of prosecution. Planning and listed building applications should follow as soon as possible, with record photographs of the condition of the fabric. The initial notice needs to explain at least the following:

- Why the works are needed urgently in the interests of health and safety and the preservation of the historic fabric
- Why the work is needed urgently rather than carrying out a temporary repair, support and shelter whilst the normal consent process is carried out
- That the works are of a limited nature and are the minimum measures needed in the interests of health, safety and the fabric

Demolition of a listed building

If it is proposed to demolish all or part of a listed building, then a listed building consent is always needed. Applications to demolish a complete listed building will be examined with great care and are only very rarely granted consent.

Since a judgment in a case between R (SAVE Britain's Heritage) v Secretary of State for Communities and Local Government [25 March 2011], any person proposing to demolish a listed building, a building in a conservation area, a scheduled monument and any building other than those which are, or adjoin a dwelling house, should consider whether a screening opinion that states that environmental impact assessment is not required, prior to making an application. It is therefore advisable to contact the local planning authority and discuss this where there is any doubt. Where demolition works are likely to have significant effects on the environment, the developer will have to apply for planning permission, and submit an environmental statement (ES) with the application.

Ecclesiastical consents

There is an ecclesiastical exemption from the listed building application process for works to places of worship for certain denominations including the Church of England, Church of Wales, the Roman Catholic Church, the Methodists Church, the Baptist Union and the United Reformed Church. All these churches run their own systems of consent which parallel the secular systems. The ecclesiastical exemption allows the needs of the 'life and mission' of the church to be considered and weighed against the needs to preserve the significant architectural and historic fabric of the place of worship. This can allow changes to churches which would be difficult to justify on a secular building if they ensure the church's continued use, and also to deal with changing numbers in the congregation. For example a major church reordering may lead to old pews and flooring being removed to allow new heating and to suit the current liturgy.

When working on a church it is always worth discussing the proposals with the church organisation who gives advice and consent. Once the architect is ready to pursue an application, a statement of need and a statement of significance need to be prepared along with an application form and accompanying specifications and drawings. The statement of need can be seen as the equivalent of a design and access statement in the secular listed building application process, whilst the statement of significance is similar to a heritage statement. Since January 2014 the applicants need to consult any relevant advisory and amenity groups such as the Victorian Society, Georgian Group, SPAB, as well as English Heritage (if the church is listed) and the local authority conservation officer if there are external alterations. The Anglican Church has a Diocesan Advisory Committee (DAC) in each diocese who advises on alterations to churches. These committees have members with expertise in the field and may include architects, clerics, archaeologists, historians, as well as representatives from heritage bodies and English Heritage, local planning authorities and the national amenity societies.

Once the architect has prepared a scheme, a statement of significance and a statement of need, and got advice from advisory groups, s/he needs to apply to the DAC for either informal or for formal advice. After considering the application and advice from advisory groups, the DAC will issue a certificate recommending the works, a certificate making no comment, or a certificate recommending a refusal. These certificates usually have certain conditions. This could be seen as being equivalent to the conservation officers recommendations to a planning committee. The final consent is called a **faculty** and is issued by a judge (normally known as a chancellor) who is appointed by the Bishop of each diocese. Once the architect has a DAC certificate, solicitors who run the Registry of Faculties for the diocese will issue notices which need to be displayed to the public and congregation prominently for 28 days. Once these notices have been displayed they are submitted to a judge appointed by the diocesan bishop for the faculty. They will consider any objections and make a ruling on the application. The Anglican system normally adheres to the following steps:

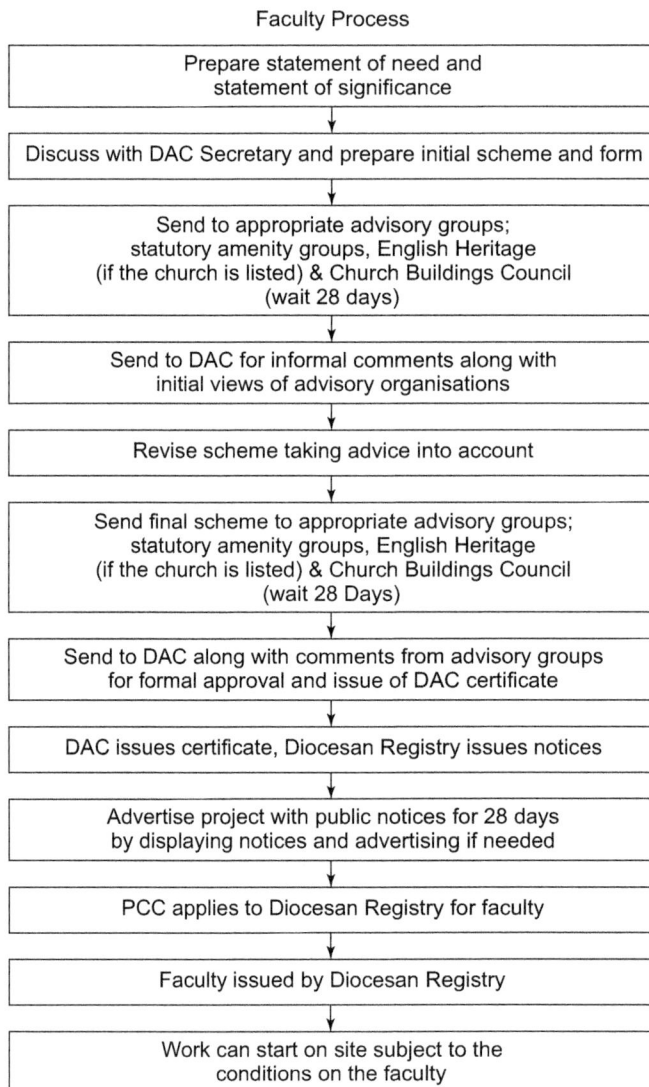

Faculty Process

Prepare statement of need and statement of significance

↓

Discuss with DAC Secretary and prepare initial scheme and form

↓

Send to appropriate advisory groups; statutory amenity groups, English Heritage (if the church is listed) & Church Buildings Council (wait 28 days)

↓

Send to DAC for informal comments along with initial views of advisory organisations

↓

Revise scheme taking advice into account

↓

Send final scheme to appropriate advisory groups; statutory amenity groups, English Heritage (if the church is listed) & Church Buildings Council (wait 28 Days)

↓

Send to DAC along with comments from advisory groups for formal approval and issue of DAC certificate

↓

DAC issues certificate, Diocesan Registry issues notices

↓

Advertise project with public notices for 28 days by displaying notices and advertising if needed

↓

PCC applies to Diocesan Registry for faculty

↓

Faculty issued by Diocesan Registry

↓

Work can start on site subject to the conditions on the faculty

FIGURE 7.2 The Church of England faculty process

Roman Catholic Church

Other denominations run similar systems with different names. The Roman Catholic Church set out their procedures in the *Directory on Ecclesiastical Exemption from Listed Building Control* (2001)[2] issued by the Bishops Conference of England and Wales. It can be found in the Bishop Conferences of England and Wales website under the Patrimony section (www.cbcew.org.uk). Applications are made to the Historic Churches Committee in each diocese using the appropriate form and accompanying documents. These diocesan Committees, like the Anglican DAC's, have advisors from English Heritage or Cadw, a representative from the local planning authority and the national amenity societies. Notices sent back to the church from the Historic Churches Committee Secretary will need to be displayed for 28 days in prominent positions near the entrance to the church. The Historic Churches Committee Secretary will also notify the local planning authority, English Heritage or Cadw and the national amenity societies. The committee is responsible for issuing the consent in the form of a faculty.

Baptist Union

Not all Baptist churches are covered by the ecclesial exemption and many strict Baptist churches are not covered. Only Baptist churches which are held in trust with one of the Baptist trust corporations and have membership of the Baptist Union operate under the rules of the exemption. Applications are made to the relevant trust. The trust then sends them onto the National Listed Building Advisory Committee, which operates a similar system to that of the Anglican or Roman Catholic church. Once the committee has consulted the heritage bodies, amenity societies and local authority, it will recommend a decision for endorsement by the trust corporation.

The United Reformed Church

The United Reformed Church publishes its procedures in a handbook produced by their network of property, legal and trust officers (PLATO) called the Property Handbook, which is found on the URC website (www.urc.org.uk). Once approval for a project has been obtained in a local church meeting and from the local district Synod Council, the church then applies to the District Synod (or provincial) property committee and their Listed Buildings Advisory Committee (LBAC) for approval. Once the Property Committee has consulted the heritage bodies, amenity societies and local authority, it will issue the decision. It is normally sensible to take the informal advice of the LBAC early in the process prior to submitting an application.

The Methodist Church

The Methodists Church has a central listed building advisory committee based in Manchester, which has a full-time conservation officer. All works have to be approved by the property committee who require that a statement of significance and need be submitted. Once the managing trustees have finalised their proposals, it is forwarded through a circuit meeting of district committees to the property committee who consult the heritage bodies, amenity societies and local authority, and advertise the proposals on site and in the local press. The Listed Building Advisory Committee will then consider any representations and make a recommendation to the property committee who issue the decision. The Church conservation officer should be consulted at an early stage and can be helpful in negotiating the process.

The ecclesiastical system only covers matters which would normally be approved with a listed building consent, and any alteration to the exterior of a listed church or cathedral will also need planning permission which is given by the local planning authority in the normal way.

Table 7.1 Works to a listed building: Summary of permissions and consents needed

Type of work	Type of Consent Required
All work of repair, alteration or demolition to a **scheduled monument**	Scheduled monument consent
Development in a **historic garden** or affecting its setting	Planning applications will be referred to English Heritage by LPA
Development in a **historic battlefield**	Planning applications will be referred to English Heritage by LPA
Total demolition of **listed building**	Listed building consent
Partial demolition of **listed building**	Listed building consent nearly always required
External alteration or extension of a **listed building** which is not permitted development	Planning permission and listed building consent
External alteration or extension of a **listed building** which would normally be permitted development on an unlisted building	Planning permission needed as listed buildings do not have permitted development rights. listed building consent almost certainly needed unless it can be shown that the character of the building will not be altered
Minor external alterations to **listed building** which is not development, such as repairs	Listed building consent normally needed except where it can be shown that the character of the building will not be changed
Alteration to interior of **listed building**	Listed building consent almost certainly needed unless the section of building to be altered has been given an exemption from listing, except where it can be shown that the character of the building will not be changed
Demolition of pre-1948 building in the curtilage of **listed building**	Listed building consent almost certainly needed unless the special character is unaffected or it is has been given an exemption from listing
Alteration or extension of pre-1948 **curtilage building**	Planning permission needed. Listed building consent may be needed where the works affect that special character of the listed building or the curtilage buildings, unless it has been given an exemption from listing and the alteration will not affect the character of the listed building
Demolition of post-1948 **curtilage building**	Listed building consent rarely needed, only where it affects the character of original Listed buildings
Alteration or extension of post-1948 **curtilage building**	Planning permission needed unless alterations covered by permitted development rights

Type of work	Type of Consent Required
Internal alterations to a post-1948 **curtilage building**	No consents needed unless there is a change of use
New building in curtilage of **listed building**	Planning permission needed
Alterations to a means of enclosure within the curtilage of a **listed building**	Planning permission required as well as listed building consent if it is attached to the listed building or will affect the character and setting of the listed building
Works to exterior of a listed place of worship of an **exempt denomination**	Planning permission needed if it can be seen (i.e. not hidden by parapets etc.) Ecclesiastical consent under the denominations own procedures will be needed if the faith denomination has ecclesiastical exemption
Works to interior of a listed place of worship of **exempt denomination**	Planning permission will not be needed. Ecclesiastical consent under the denominations own procedures will be needed
Works to a listed place of worship of **non-exempt denomination**	Treat as an ordinary listed building, and obtain listed and planning consents as necessary

Changes to the Heritage Asset Consent Regime 2014

The government have made some changes to the heritage asset consent regime to simplify the planning system following the Penfold Review of non-planning consents 2010 and the Enterprise and Regulatory Reform Act 2013. The measures came into force in 2013 and 2014 are as follows:

* Amendments to the Planning (Listed Buildings and Conservation Areas) Act 1990 will allow structures or objects attached to or within the curtilage of listed buildings to be excluded from the listing, and for specific parts or features of such buildings to be identified as lacking special architectural or historic interest.
* The regime for certificates of immunity from listing is to be widened to allow them to be issued at any time rather than just following an application for or the grant of planning permission.
* The act also introduced Local and National Heritage Partnership Agreements which enables works to be carried out over time without repeated applications for listed buildings consent.
* The act removed the need for conservation area consent for the demolition of unlisted buildings in conservation areas, leaving such matters to be dealt with through the grant of planning permission.
* There are also provisions relating to enforcement in connection with the demolition of buildings.

Making the listed building consent application

The process of making a listed building application is similar to that made for a planning application. As John Collins and Philip Moren explain in *Good Practice Guide to Negotiating the Planning Maze* 'the key to successful application will often lie in its careful preparation and presentation'[3]. This is especially true for listed building applications where the success of the application will often depend on the conveying of a clear understanding of the significance and history of the building and the reasons why the proposals will not harm the character of the significant fabric.

Many applications fail to be registered because vital elements of information are missing. Councils produce validation check-lists which are meant to cover all eventualities, and these can be very helpful when considering the information that needs to be provided. There are now a wide range of statements and assessments which can be requested when making a listed building application.

Mandatory requirements for listed building applications

Guidance can be found on the Planning Portal (www.planningportal.gov.uk) into the documents needed to make a valid planning application. Applications for listed building consent must include the following:

1. The standard listed building application form with the ownership certificate signed.
2. A location plan based on a current map to a scale of 1:1250 or 1:2500. This plan should wherever possible show at least two named roads as well as naming the neighbouring properties. The application site should be outlined in red and include all the land necessary to carry-out the works. This would normally be done round the curtilage of the listed building. A blue line should also be drawn around any other land owned by the applicant close to or adjoining the site. There should be a scale and a north point marked on this plan.
3. There should be a site layout/block plan which is normally drawn to a scale of 1:200 or 1:500. This plan should show:
   - A north point
   - The proposed works location and the position of any buildings in the curtilage of the listed building
   - A scale marker so if printed from a website the scale can be worked out
4. All listed building applications require a design and access statement even when they are for a client who is a householder (where one would not be needed for a planning application). The design and access statement should address at least the following:
   - A written statement that includes a schedule of the proposed works
   - An explanation of and justifications for the proposals

The design and access statement is normally expanded or supplemented by a detailed heritage statement. As explained in Chapter 5 the main purpose of the heritage statement is to convey your knowledge about the significance of the fabric. The amount of information required for a listed building application will vary with the circumstances of each application, and the scope of a heritage statement may need to be expanded and agreed with the conservation officer before submission. It may include the following additional information:

- A structural survey of the fabric detailing repairs and changes needed to the fabric. This would normally be done by a structural engineer and include his/her justification for the type of repair and how it limits its impact on the original fabric

- Where the works involve disturbing the ground, an archaeological scoping study assessing the impact on any likely archaeology and mitigation measures is required, such as recording archaeological digging or archaeological watching
- Where works involved may affect the setting of a listed building, garden, historic park or historic battlefield, a written statement may be required detailing the impact of the proposals on the special character of the historic features
- Specialist advice from experts in the conservation of specialist elements of the fabric

Depending on the nature of the application it will be necessary to include the following drawings, which should always include a north point and a scale marker on them for digital viewing:

- **Existing and proposed elevations** at a scale of at least 1:100 or more likely at 1:50 or even larger. The scale will need to be appropriate to show the architectural details being altered. For example individual timber studs or the position and thickness of lath and plaster may need to be shown, which is difficult often below a scale of 1:20. Any external alterations will need the elevations to be drawn and show the proposed building materials, the style and finish of joinery including doors and windows. Where the building adjoins another building the relationship between the buildings need to be clearly shown. Where walls or other features are to be demolished they should be clearly shown and on more complicated applications it can be helpful if a separate drawing showing all the demolitions is produced.
- **Roof plans** at a scale of at least 1:100 or more likely at 1:50 or larger. These will be needed where changes are proposed to roof shape, roofing materials and drainage. Materials should be shown on these plans.
- **Existing and proposed sections and site levels** at a scale of 1:100 or more likely at 1:50 or 1:20 or even larger detail. These drawings should be produced to show any changes of finished level which are proposed inside or outside the building. Level changes can affect the character of the building and also have an impact on archaeology hidden below the existing surfaces. They can also show the heights and thickness of internal features.
- **Room elevations, ceiling and floor finish plans** at a scale of 1:20 or 1:10. Where alterations are made to the interior of a listed building it may be necessary to draw room elevations to show repairs and alterations in detail. Unlike an unlisted building, internal alterations may need consent and be described. Considerably detailed drawings of rooms are often needed to illustrate the proposals.
- **Details of construction** at a scale of 1:20, 1:10, or a larger scale. Often when altering a listed building, details are needed of the joinery as well as alterations to the construction of the building. The character of a listed building can be changed by poor detailing, therefore detailed considerations of elements of construction can be required to be agreed with the council. Often the principle of the changes to the buildings fabric can be agreed with the initial listed building application, which is then granted subject to final joinery details at a large scale being submitted for approval.

When replacing or adding windows and doors the correct detail of mouldings, etc can only be described by drawing a detailed drawing with sections of mullions, glazing bars and jambs to a scale of 1:5 or larger.

When converting a historic building the requirements of the building regulations can require roofs to be lifted to fit in insulation or historic walls lined, and these changes will need detailed design and consideration prior to listed building consent being granted, as the changes will affect the character of the historic building. Conservation officers will often require an application to be at the level needed to submit a building regulations application. This is far more detail than normally needed for a planning application.

- Photographs
  The recording of the listed building prior to any works is a good conservation practice. Recording helps to ensure historic records are available for future conservators, architects or historians. Photographic surveys of a building prior to changes can be a very useful record for the local planning authority considering the application, and by including a good and comprehensive set of digital photos in an application will ensure there is an enduring record of the building before the changes. When major changes are made to historically significant buildings such as ancient churches, record photographs are often lodged in the diocesan or county records archive or even the national archives as a condition of consent.
  Photographs supplemented by notes or shading can be a very efficient way of identifying the location of specific repairs, and annotated photographs can often be used in lieu of drawings for applications for repair works.
  Laser cloud surveys are a new technology where a survey is carried out in three dimensions using millions of laser beams sent out from a special instrument. These cloud surveys can produce outputs which look like a colour photographs, but they are extremely accurate and can be measured just like a drawing. The surveys can also be used to build accurate 3D CAD models and are currently the most accurate and complete form of recording available.

Making an application for a non-designated heritage asset

When making a planning application involving an older building which may be considered to be a non-designated heritage asset by the local planning authority, it is still a requirement to consider if it has any heritage significance. There is no need to write a separate heritage statement, however paragraph 128 of the NPPF requires information about the significance of the fabric and the impact of the changes to be included in the design and access statement (normally under the heading of 'heritage assessment') including:

- An analysis of the architectural, historical, aesthetic or cultural significance of the building
- The way an architect's proposals will affect the significant features of the building
- This document would normally include photographs of the significant architectural features affected by the proposals to make it easier for the conservation officer and planning committee members to understand the proposals
- The building's setting and how it will be affected by the proposed works

Fees

There are no fees for listed building applications. It is hoped that because there is no fee for applications on listed buildings, owners will not be discouraged from making an application and there will be no incentive to them to flout the law and carry out work without consent. It is a criminal offence to do work to a listed building without approval, which is subject to a maximum penalty of imprisonment for two years and/or an unlimited fine. However, this alone has not been felt to be a sufficient incentive for owners to comply with the law. In an age where local councils are under severe financial pressure and many conservation departments are under staffed, there are calls for fees to be imposed on listed buildings and therefore this may change in the near future.

How the council processes an application

The application process in England is meant to take 8–13 weeks from registration, depending on the complexity of the application.

Local authorities have a duty to consult the local regional office of English Heritage as soon as they receive an application for:

- Any works to a Grade I or II* building
- The demolition of a Grade II building
- The demolition of the principal external wall of a Grade II building
- The demolition or substantial alteration to a Grade II listed building

English Heritage will then offer help and support from its experts on the application, which can have considerable influence on the outcome.

Local authorities also have a duty to notify the major national amenity societies of any application to demolish or significantly alter part or whole of a listed building, no matter what grade the building is. These bodies act as expert advisors and guardians of our heritage. Normally the conservation officer will advise on when they need to be consulted or do it him/herself during the application process. On major applications however, it is often worth consulting these bodies prior to the application to ensure there are no delays in waiting for their responses. These amenity societies are charitable bodies with a large case-load and generally very meagre resources, and it can take time for their hard pressed case officers to respond to councils. They are meant to reply to the local authority within 28 days, but often their workload and committee dates make this impractical. In practice most planning departments are reasonably tolerant to the societies and consider responses received late. The National Amenity societies which are consulted are:

- **Ancient Monuments Society**, which advises on the future of the country's ancient monuments, including pre-medieval fabric including Roman. They are unique amongst the national amenity societies in that buildings of all ages fall within their remit. 'Ancient monument' is an ambiguous term, which they interpret as any man-made structure of architectural or historical interest – including houses, whether vernacular or polite, barns, alms houses, dovecotes, mills, churches and chapels.
- The **Council for British Archaeology (CBA)** advises on all matters where archaeologically significant fabric above or below the ground may be affected by a proposed development. Their expertise encompasses archaeological sites as well as ancient and historic buildings, monuments, and antiquities.

- **The Society for the Preservation of Ancient Buildings (SPAB)** concentrate their casework on significant applications to alter or demolish buildings constructed before 1720
- **Georgian Group** concentrates their advice on fabric and structures and gardens created in the Georgian period (broadly 1700–1837)
- **Victorian Society** is concerned with the preservation of Victorian and Edwardian Architecture from 1837 until the end of the Edwardian period, which is seen as either the death of King Edward VII or the start the First World War in 1914
- **20th Century Society** seeks to preserve the best of architectural heritage since 1914. Many important buildings of the twentieth century are not yet listed, and much of their casework involves making well-researched and detailed arguments to English Heritage, CADW or Historic Scotland.

These national amenity societies also have to be consulted by the applicants or religious bodies administering the ecclesiastical exemption including diocesan advisory committees of the Church of England. These bodies write to them informing them of any applications for demolition or partial demolition of a listed church, and again as with secular buildings, they normally give them at least 28 days to respond.

In addition to the National Amenity Societies, the Diocesan Advisory Committees of the Church of England normally need to consult the Church Buildings Council on alterations which involve significant intervention in Grade I and Grade II* listed churches, as well as for all larger significant interventions and proposals which would have high or moderate impact on a church building or its interior or setting. They would also be expected to be consulted on controversial cases and where there is no generally agreed solution from the advisory bodies, amenity societies and parishioners. The Church Building Council is a statutory body and not an amenity society, whose role is in support of furthering the mission of the Church of England.

Discharging listed building conditions

At the end of the planning process a listed building consent is hopefully granted. It is usual to find a string of conditions on a listed building consent which require further work, and it is a rare event for works to a listed building to remain totally unaltered by discoveries made on site as work proceeds. Typical conditions needing additional information to be discharged include:

- Samples of materials or mortars may need to be approved by a conservation officer
- Final details of joinery may need to be submitted and approved
- Specific investigations on site may need be carried out and inspected by the council
- There may be a need for an archaeological investigation prior to work starting
- Archaeologists may need a 'watching brief' where they are present, and recording whilst excavations are made. If significant archaeology is found, plans may need to be altered to avoid it
- Approval for alterations to original consent agreed in the light of on-site research. Often variations can be agreed by letter with the local planning authority, but some may require additional listed building applications

Appeals

Inevitably some applications are refused or a local planning authority fails to make a decision within the statutory period. These could be because there are onerous conditions placed on the applicant, leaving the applicant dissatisfied or aggrieved. In these circumstances the applicant has the choice of revising their proposals and re-submitting, or using their statutory right to appeal to the Secretary of State (or equivalent) against the decision of the authority.

To appeal to the Secretary of State, one needs to apply within three months of the receipt of the decision notice using the appropriate form if the application is for a householder, and six months for all other applications. In England and Wales appeals are sent to the Planning Inspectorate in Cardiff or Bristol, in Northern Ireland they are sent to the Planning Appeals Commission in Belfast, and in Scotland they are sent to Scottish Executive Inquiry Reports Unit in Falkirk.

The appeal application should include:

- The appropriate form which can be downloaded from the internet or completed online at the Planning Portal website on (www.planningportal.gov.uk). On the form the applicant will need to state the grounds for the appeal
- The original application, including all supporting material including forms, drawings, reports, statements and the plan showing the site outlined in red etc
- The local planning authority's decision notice
- Any revised drawings or additional material to support the appeal

Often on listed buildings the grounds for the appeal will be because there is a disagreement between the applicant and the local authority over the relative significance of a piece of fabric, which may make it necessary to support the applicant's views with independent expert analysis.

Procedures

Normally it is possible to choose to make an appeal by an exchange of written representations or through a public inquiry or informal hearing. Most applicants opt to make the appeal by written representations as it is much less expensive to mount and normally quicker. The applicant may well need the services of a specialist town planning consultant to help with an appeal using written representations or at an informal hearing, whereas they will probably need the skills of both a planning consultant and a solicitor or barrister at a public inquiry.

On receiving the appeal application the inspectorate will write to the parties and determine a 'starting date' for the process. The local planning authority have to then advertise the appeal and give notice to any interested third parties such as neighbours or objectors to the original application. All parties are given six weeks to submit additional representations and then each side has a chance to comment on the representations. The inspector may visit the site accompanied by representatives of both parties, after which the inspector will write up and issue his decision.

An informal hearing follows a similar timetable, but instead of representations, each side submit hearing statements together with supportive documents within six weeks of the 'starting date', and written comments on the other parties' statements within nine weeks. In theory the hearing is held within 12 weeks but often this is delayed. The hearing process is informal and led by the inspector and does not involve any cross examination.

Public inquiries need longer to prepare for and can last for many weeks. They are normally used on the most complicated and high profile cases and do involve cross examination of witnesses, which is why specialist solicitors and barristers are normally used. The Planning Portal (www.planningportal.gov.uk) publishes information regularly about the length of time appeals are taking from the starting point to determination.

Notes

1   HM Government (1979), Planning (Listed Buildings and Conservation Areas) Act 1990, S7
2   Bishops' Conference of England and Wales (2001), *Directory on Ecclesiastical Exemption from Listed Building Control* (2001), The Catholic Church in England and Wales, available from www.cbcew.org.uk
3   Collins, J. & Moren, P. (2009), *Negotiating the Planning Maze*, London, RIBA Publishing, p. 70

Further reading

> Collins, J. & Moren, P. (2009), *Negotiating the Planning Maze*, London, RIBA Publishing.
> HM Government (1979), *Planning (Listed Buildings and Conservation Areas) Act 1990*, available from www.legislation.gov.uk
> HM Government (1979), *Enterprise and Regulatory Reform Act 2013*, available from www.legislation.gov.uk
> Mynors, C. (2006), *Listed Buildings, Conservation Areas and Monuments (Fourth Edition)*, London, Sweet & Maxwell.
> Taylor, J. (2010), *Listed Places of Worship, Building Conservation Directory 2010*, Tidsbury, Cathedral Communications Ltd, also available at www.buildingconservation.com
> Church Buildings Council, *Various Guidance Notes on making changes to Anglican Churches*, London, Church Buildings Council available from www.churchcare.com
> www.cbcew.org.uk
> Cadw
> www.urc.org.uk
> Enterprise and Regulatory Reform Act 2013
> www.planningportal.gov.uk

CHAPTER 8
RECONCILING CONSERVATION NEEDS AND BUILDING REGULATIONS

Listed buildings still need approval for alterations and structural repairs under building regulations in the normal way. The regulations apply both to new and existing buildings and they also apply to listed buildings and heritage assets which are occupied. Sometimes however, it can seem impossible to reconcile the requirements of the conservation department to preserve significant fabric and those from building control and in these cases it is often necessary to obtain waivers or derogations (partial exemptions) from normal building regulations standards, with the aid of the conservation officer and building control officer. Negotiations often are needed to balance the competing needs of preserving the character of the historic fabric and the regulations. It is good practice to seek an early consultation between these officers at the listed building application stage to establish the scope for improvements without affecting the building's character. Once they are established, it is possible to go on to make the listed building application with the certainty that the applicant's proposals will also get Building Regulation Approval. English Heritage has published useful advice in an interim guidance note which is available on their website and called 'Buildings Regulations and Historic Buildings' although this document has not yet been revised following recent changes in the building regulations.

Part B – Fire safety

The building regulations Approved Document B (Fire Safety) allows for variation of standard provisions in historic listed buildings where it would otherwise be restrictive and damaging to the character of the listed building. However, where a client identifies risks in complying with their duties under the Regulatory Reform (Fire Safety) Order 2005, or there is a material change of use, a significant alteration, or extension, there is still a need to take measures to meet any site-specific risks and hazards. In these cases it is often sensible to use fire engineering techniques, which may include smoke modelling and building fire performance analysis, to make a site-specific risk assessment. Following the assessment it should be possible to mitigate the specific risks. In historic buildings it is often best to use fire safety management systems which have little impact of the character of the building, such as automatic detection systems, suppression systems and fire safety management. The alternative approach often requires the making of physical changes, such as new escape doorways and upgrading the fire resistance of walls and doors, which are likely to change the character of the building and impact on its significance.

Part E – Sound transmission

When making certain changes to the material use of a building, Part E of the building regulations often requires that the existing structure is upgraded to reduce the passage of sound, say from one flat to another. The approved document accompanying the building regulations does accept that this is not always practical in historic buildings and in these cases derogation from the normal standards is normally sought. This improves as much as possible the resistance to sound transmission without damaging the significance of the fabric or risking long term decay.

Part F – Ventilation

Historic buildings are often well ventilated and contain flues, windows and doors which are not fully airtight, so in practice, meeting the provisions of Part F of the building regulations to provide sufficient ventilation for the building occupants is rarely an issue. However when new windows are fitted, or there is a material change of use, the Approved Document F (Ventilation) gives guidance on fitting trickle vents, and for the need for windows to provide rapid purge ventilation and under certain circumstances to be sufficiently large to provide a means of escape. It is normally accepted that if the original listed building windows did not have a trickle vent it need not be included in its replacement. However the need for purge ventilation and having windows of sufficient size to provide a means of escape are both issues to improve the safety of the occupants, and building control officers will often be reluctant to agree derogations for these items without alternative safety features such as sprinkler systems being in place.

Part L

Conflicts between the need to preserve the architectural and historic character of the listed building and the requirements of the building regulations most commonly arise when the requirements of Part L of the regulations dealing with the conservation of fuel and power are strictly enforced. For example, the conversion of an old mill to a dwelling should comply with the requirements of Part L for thermal insulation of the external envelope, but the lining of walls with insulation and changing single glazed windows for double glazed windows may destroy the character of the building and thus may be unacceptable to the conservation officer. Also sealing the roof of the mill and adding modern impermeable insulation may lead to condensation forming around historic timbers which will lead to deterioration of the fabric due to rot.

Where improvements to historic and traditional buildings trigger the need for compliance under Part L of the Building Regulations the guidance in the Approved Documents L1B Conservation of Fuel and Power in existing Dwellings and Approved Documents L2B Conservation of Fuel and Power in Buildings Other Than Dwellings both acknowledge, in their respective Clauses 3.8–3.13 that 'special consideration in making reasonable provision for the conservation of fuel and power" can be accommodated. For these buildings "the aim should be to improve energy efficiency as far as reasonably practical. The works should not prejudice the character of the host building or increase the risk of long term deterioration of the building fabric or fittings'[1].

Definitions of a historic or traditional building in Part L

Building Control should agree to derogation from the normal standard if the building has a statutory definition of being an historic building and to comply with the regulations would unacceptably altar the character of the building. However the Approved Documents L1B and L2B in paragraphs 3.7 and 3.8 also agree special considerations for some other classes of historic structures which include:

[a] Listed Buildings
[b] Buildings Situated in a Conservation Area
[c] Scheduled Ancient Monuments
[d] Buildings which are of local architectural interest and which are referred to as a material consideration in a local Authority's development plan (i.e. on a local List)
[e] Buildings of architectural and historic interest within National Parks, Areas of Outstanding Natural Beauty and World Heritage Sites
[f] Buildings of traditional construction with permeable fabric that both absorbs and readily allows evaporation of moisture.

In Conservation Areas, National Parks, World Heritage sites and Areas of Outstanding Natural Beauty and for buildings on a local list, only changes required under Part L of the building regulations to the external appearance can get a derogation from the building regulations. For example the architect is unlikely to get consent to replace traditional single glazed windows for more thermally efficient double glazed windows made of non-vernacular materials such as plastic.

There are other buildings that fall outside any form of legal protection which have historic features that contribute to the character of an area, including buildings in historic parks and gardens, buildings within the curtilage of a scheduled ancient monument, or fine examples of local vernaculars where changes to the external appearance, especially the windows, will harm the building's appearance. In these cases sometimes it is possible to appeal to a conservation officer in justifying resisting the changes wanted by building control. There is no statutory requirement for the building control officer to take the advice of the conservation officer but sometimes they do. In very rare cases if the conservation officer feels strongly that the changes to the external appearance are not justified s/he has the final sanction of nominating the building to be added to a local list and using an Article 4 direction to prevent the work.

Special characteristics of a historic building

Before approaching building control for derogation from the normal regulations, it is useful to be clear in defining the special characteristics of a building. As emphasised elsewhere in this book the key to good practice, when working on a historic building is to have considered the architectural and historic significance of the building prior to proposing any work, and having done this, conveying the findings to building control and the conservation officer. If the architect is making a listed building application they will have articulated the special elements in a heritage statement or a statement of significance. Typically the features the architect would be discussing that could be affected by the building regulations would include:

- **External coverings** – including roof tiles, bricks and renders etc. Changes to the external materials can clearly have a marked effect on the special character of a building. Rendered buildings can have their thermal performance improved by having the render fixed onto insulation boards, or by using special insulating renders instead of more traditional ones. However, these techniques will increase the thickness of the wall coverings and this can change the character of the building by reducing eaves overhangs and increasing the depth of reveals. They can also prevent the existing fabric breathing and letting out trapped moisture which can encourage decay.

- **Windows and doors** – replacing traditional joinery including single glazed windows with thermally broken, double or triple glazed windows is often part of the energy efficiency measures used in refurbishing a building, but it will significantly change the character of a historic building. The character of a historic elevation is normally formed by the window and door openings, their proportions and their subdivisions, therefore changes to these elements is normally considered by both the Conservation officers and building control officer to be unacceptable. Traditional ways to reduce heat loss from windows included shutting heavy curtains and closing shutters, and these methods are very effective at night. Nowadays acceptable ways of improving the thermal efficiency of windows and doors in the daytime include the use of secondary glazing and application of draught proofing. There are now some very thin double glazed units which sometimes can be used within existing frames but even these can often have a noticeable effect on the character of the window. It should also be remembered that the old glass in original windows is often historically important and can give interest to an elevation because they reflect less uniformly than modern float glass.

- **Internal coverings** – internal walls can be covered with timber panelling or historically significant materials like wattle and daub or lath and plaster. Changing these for insulated dry-linings in a listed building will change its special character and therefore not be acceptable to the council; it may however be acceptable on an unlisted building in a conservation area as it would have no impact on the external appearance. Lining walls will also have an impact on cornices, skirting boards, door and window linings – all of which give a room its internal character. Thermal linings can also prevent the existing fabric breathing and letting out trapped moisture which can encourage decay.

- **Chimney pots** – changing a boiler for a more efficient one can require that a chimney pot needs to be changed, but the style of chimney pot – especially if it is large or decorative – can contribute significantly to the character of the building.

- **Vent pipes** – new plumbing and ventilation inside a building often needs new ventilation pipes and extract grills. The placing of these can be sensitive and disrupt the traditional character of a wall or roof slope. The use of air admittance vents on soil stacks can reduce the need to have vents going through roofs, and discrete ventilating tiles, slates and ridge vents can disguise them. Extract vents can be also partially be hidden in eaves.

- **Floors** – in listed buildings, old, worn, uneven floors may be part of the special character of the building and therefore they should not generally be lifted. However there are circumstances when they can be lifted, which may give the opportunity to install flooring insulation beneath the floor before putting them back. Suspended timber floors are also common in eighteenth and nineteenth century buildings, which can be relatively easily insulated between the joists, although it is essential to ensure the void beneath is well ventilated to prevent condensation.

- **Roofs** – Traditional roofs are normally well ventilated and this helps dissipate any moisture which quickly rises into the void. Traditional roofs relied on the coverings, be it slates or tiles or even thatch, stopping rainwater getting in. These were often far from vapour tight and allowed air to blow through them. If maintained, these roofs are very effective at keeping roof structures dry but where building control insist on the introduction of a membrane under the covering it is best, in historic buildings, to use one that is vapour permeable. Any membrane including a vapour permeable membrane will reduce the air flow in the roof void and it may also be necessary to add additional ridge or eaves ventilation to ensure the roof stays well ventilated.
  Care also needs to be taken when improving smaller roof voids, say beneath a flat roof or roofs, where the rafters are internally boarded or have plaster linings. To ensure these smaller voids are ventilated, especially when introducing insulation, it may also be necessary to include a vapour control layer on the warm side of the roof to prevent excess moisture entering a small, less easily ventilated space.

- **Heating systems** installations in historic buildings can also be sensitive and have a significant impact on the special character of the building. Considerations which may need to be addressed include:
  — What temperature can the fabric tolerate? Older historic buildings will never have experienced high heating levels, and if this is done, the interior fittings may warp or crack. For example the Church Buildings Council recommends that medieval churches are only heated to a maximum of 18°C+ or −1°C to protect medieval timbers and organs
  — What will be the heating source? The use of low temperature heat sources such as low temperature radiators or underfloor heating can reduce the impact of the heating gradients on the surrounding historic fabric
  — The way heating sources and distribution services interrupt important and attractive internal views needs to be considered
  — The need to make holes in historic fabric for pipes or flues need to be minimised or preferably eliminated
  — Proximity of uninsulated pipes and heat sources such as radiators to important fabric such as panelling which can be damaged by overheating and drying out
  — Placing radiators on historic walls can also encourage moisture to wick up through the porous vernacular structures behind them. This can lead to alterations to the moisture balance in the wall and cause crystallisation of salts in the wall at the surface leading to deterioration of the finishes and plasters

- **Lighting and power systems** – the installation of new lighting systems can also affect the special character of the building. The location of new fittings and cable runs needs to be done with care to avoid unnecessary holes in the fabric and disturbing the internal ambience of the interior.

Statutory required upgrades

There are certain upgrades to conserve fuel and power that any building control officer will expect to be made when altering or extending, even when the building is a designated historic asset. In L2 these items include:

- Making reasonable efforts to limit heat loss and gain through the fabric
- Limit the heat loss of hot water pipes and hot air ducts by insulating them
- Limit the heat loss from hot water tanks etc by insulating them
- Provide efficient hot water and space heating devices
- Make reasonable provision to limit solar shading
- Provide sufficient information about the relevant services so the building can be operated and maintained in a manner as to use no more energy than is reasonably necessary

In larger buildings and non-domestic buildings the regulations require:

- Air conditioning and mechanical systems to be installed to minimise their energy demand to be no more than reasonable in the circumstances (buildings over 200m^2)
- Limit the heat gains of chilled water and refrigerant vessels, ducts and pipes (buildings over 200m^2)
- Provide energy efficient lighting systems

Part M – Access to and use of buildings

When a material change is made to a Listed Building, such as a change of use, or an extension is added, then there is a need to provide reasonable levels of access for all groups including the frail and disabled. The Equality Act 2010 also requires service providers to take reasonable steps to remove physical features which impede access to disabled people. It is generally accepted that the aim should be to provide better access without damaging the significance of the listed building or increasing the long term risk of deterioration. For example when making provision for disabled access into historic churches, schemes which often get consent have had to compromise normal standards of disabled access in order to preserve significant historic fabric. English Heritage have produced a series of guides to equality of access to listed buildings and other Heritage Access, including one for historic buildings and one for historic landscapes[2].

Part N – Glazing

Older windows are often hard to clean safely and made with thinner glass than would be used today. Part N of the building regulations covers the requirements for safety in relation to impact, opening and cleaning of glazing for new windows, however these requirements do not apply to the installation of replacement windows or glazing repairs to listed buildings unless there is a material change of use of the building.

Notes

1   HM Government (2010), *The Building Regulations: Approved Document L1B (Conservation of Fuel and Power in Existing Buildings Other than Dwellings)*, Cl. 3.9, p. 9 and
HM Government (2010), *The Building Regulations: Approved Document L2B (Conservation of Fuel and Power in Existing Buildings Other than Dwellings)*, Cl. 3.9, p. 10

2   English Heritage (2013), *Easy Access to Historic Landscapes*, London, English Heritage, also available from www.english-heritage.org.uk

Further reading

> HM Government (2013), *The Building Regulations: Approved Document B (Fire Safety) (incorporating 2010 & 2013 amendments)*, available from www.planningportal.gov.uk
> HM Government (2010), *The Building Regulations: Approved Document E (Resistance to Sound) (incorporating 2010 & 2013 amendments)*, available from www.planningportal.gov.uk
> HM Government (2010), *The Building Regulations: Approved Document F (Ventilation) (incorporating 2010 & 2013 amendments)*, available from www.planningportal.gov.uk
> HM Government (2010), *The Building Regulations: Approved Document L1B (Conservation of Fuel and Power in Existing Buildings Other than Dwellings)*, available from www.planningportal.gov.uk
> HM Government (2010), *The Building Regulations: Approved Document L2B (Conservation of Fuel and Power in Existing Buildings Other than Dwellings)*, available from www.planningportal.gov.uk
> HM Government (2013), *The Building Regulations: Approved Document M (Access to and use of Buildings)*, available from www.planningportal.gov.uk
> HM Government (2010), *The Building Regulations: Approved Document N (Glazing – Safety in relation to impact, opening and cleaning) (incorporating 2013 amendments)*, available from www.planningportal.gov.uk
> English Heritage (2013), *Conservation Basics: Practical Building Conservation*, Farnham, Ashgate Publishing Ltd.
> English Heritage (2013), *Easy Access to Historic Landscapes*, London, English Heritage, also available from www.english-heritage.org.uk
> English Heritage (2012), *Easy Access to Historic Buildings*, London, English Heritage, also available from www.english-heritage.org.uk
> English Heritage (2011), *Energy Efficiency and Historic Buildings – Application of Part L of the Building Regulations to Historic and Traditionally Constructed Buildings*, London, English Heritage, also available from www.english-heritage.org.uk
> Building Regulations Approved Document B (Fire Safety)

CHAPTER 9
PRACTICAL MATTERS: FROM FEES TO SITE INSPECTIONS

Risk management is the key to successful conservation projects

A conservation and refurbishment project is often a process of discovery and reaction, which requires careful management from the architect and contract administrator to avoid difficulties. An experienced architect can plan the contract in ways to avoid the considerable risks of cost and time over-runs that are the result of not being able to fully establish the condition of the existing fabric before work starts. It is often impossible to accurately predict the state of the walls behind renders, joists below floor-boards, rafters and battens below tiles, but with experience it may be possible to predict, and allow sufficient time and funds to deal with the defects that are uncovered. The repair solutions to these hidden problems will often need consultation with other members of a professional team, and approval of the authorities or grant aiding bodies. Whilst these approvals are being sought the contractor may be delayed. All too often the unprepared architect finds that these variations lead to the scope of work growing significantly beyond the original budget and time needed to complete the project. Unexpected problems will often lead to unhappy clients, and therefore it is good practice in preparing contract documents to identify the likely risks and mitigate them by building in sufficient time and money to overcome any that arise. A good contract administrator therefore, explains the risks to the client and arms themselves with high quality documentation, which can be manipulated to overcome the risks that will arise during the construction period. They will also charge a fee for their services that is sufficient for them to resource the project fully.

This chapter looks at the key practical issues which are faced when administering conservation and refurbishment contracts as opposed to a new build contract. The items are ordered so they follow the process for such a commission, including:

1. Scope of work and conservation philosophy to be adopted
2. Getting the fee right
3. Budgets and initial assessments
4. Investigative contracts
5. Ecological studies
6. Asbestos
7. Structural engineers and other consultants
8. Quantity surveyors and cost control
9. Choice of contract for conservation projects
10. The form of contract

1. Scope of the work and conservation philosophy to be adopted

Most historic buildings will need some form of continued maintenance into the future, thus it is rare that a building is completely refurbished in one contract to a point when no more work will be needed to maintain the building for a significant length of time. Historic buildings, be they medieval churches or listed houses are rarely in a perfect state of repair, as most clients cannot afford to do everything at once, and decay will be carrying on at different rates to different pieces of fabric all the time. For example, if all the decaying stones of a church, or decaying bricks to house, were replaced all at once, no matter how far the deterioration had gone, you would destroy the character and patina that comes with the ageing process.

It is important at the start of a conservation project to consider the objectives to be achieved, and to match the strategy to the budget of the client. One consideration will be how much can the client afford, and which repairs are most urgent. Normally repairs that prevent further deterioration of the fabric and keep the building wind and weather tight, or prevent structural failure should take priority. There is little point repairing the panelling inside the Georgian house if the valley gutters above are left leaking water down onto them, encouraging beetle and dry rot attack. Similarly repairing the decayed stone tracery at the centre of a medieval window without repairing the hood moulding over it would not be advisable, as the hood moulding is there to shed water away from the more delicate tracery below, and without it, the new stone will continue to decay at an accelerated rate.

Another consideration will be the anticipated length of time until the next major refurbishment. Masonry on a church tower or spire is very expensive to reach, meaning that a high proportion of the repair contract on a spire will be the cost of the scaffolding rather than the repair itself. Normally such repairs are only done once every 50–100 years when there is a sufficient quantity of repairs to justify the scaffolding costs. The state of decay of some stone elements will therefore often be quite advanced. Had they been lower down, the repair may well have been done much sooner. Therefore any short term repairs with cheap stone, thin stone indents, or mortar repairs in lieu of stone, would be inadvisable. Repairs with substantial stones capable of being left unattended until the next time the spire is covered in a scaffold would be far more sensible.

Where the budget is tight and the repair is easily carried out, it may be sensible to carry out a cheaper, temporary form of repair, leaving as much historic fabric intact as possible and accepting that the repair may need to be carried out again in a few years time. For example on a church where the medieval windows are easily reached from the ground, it may be sensible to carry out a soft mortar repair (sometimes called a 'plastic repair') to fill gaps in the stone tracery so it is weathered, and paint the stone in a sacrificial lime shelter coat to help stabilise the stone and slow the decay process. These repairs are inexpensive compared to replacing the carved stonework and they have the advantage of leaving the historic fabric in place.

Historic building owners will need to have these issues discussed with them, and therefore the architect or surveyor should agree the conservation philosophy with his client. The choices made will have an effect on the future maintenance costs for the clients and they have a right to understand the implications of the choices made and any future expenditure needed on their property.

2. Getting the fee right

Architects used to the construction of new buildings often find refurbishment and conservation projects far more complex and time consuming than normal 'new build' contracts, which needs to be reflected in the fees charged. The need for the architect to understand the business costs of taking conservation work can be especially important when considering an alteration and refurbishment project. The project may have a lengthy process to reach listed building consent, and therefore it is likely to need close supervision. Also, the administration time is likely to be lengthy due to the quantity of necessary variations issued to cover work identified only after the fabric is opened up. Indeed many well-known specialist conservation architects will only work on a time basis, because of the uncertainties. Most clients however insist on ensuring the professional teams' fees are controlled, with reliance solely on time charge fees not being an option available to most architects.

It is often very difficult to establish the full scope of works on a refurbishment project until the project gets underway, and therefore traditionally, many architects relied on a percentage fee which reflected the final cost of the works. Since the removal of the RIBA mandatory fees scales used in the mid-twentieth century, competitive fee bidding and tightening profit margins have made the reliance on percentage based fees without an understanding of the likely project costs questionable. Historic buildings vary greatly in complexity and this affects the length of time needed by the architect to provide his services. It is therefore recommended that on conservation projects the architect should prepare resource based fee bids, and that time is spent properly assessing and presenting the fee needed to meet the specific requirements of the project. The RIBA has developed an Architects' Fees Toolkit and calculator to help practices prepare fee bids in an open way. The fee and service can then be defined, explained and agreed with the client, and can form the basis of a transparent and successful relationship.

Where a client insists on lump sum fees, it is advisable to carry out initial feasibility studies on a time basis. During this period, initial site investigations can be made to investigate the fabric and to establish what needs to be repaired and how much is it likely to cost. After the scope of the works is established and the budget is set, the architect should be able to develop a detailed resource model, reliably identifying the resources needed to deliver their architectural services, to a point where they can offer a lump sum fee.

The RIBA *Good Practice Guide on Fee Management* contains specific advice to help an architect get the fee right so s/he can make a profit as well as deliver an adequate service to the client.

The RIBA standard Fee Agreements has two specialist component modules which are designed to identify the schedules of services an architect provides when working on historic buildings. They are:

- Historic Building or Conservation Project Services Schedule
- Small Historic Building or Conservation Project Services Schedule

The RIBA also produces two guides for clients using these fee agreements, which are often very useful ways to explain the normal services of an architect when working on historic buildings. See Further Reading.

3. Budgets and initial assessments

When dealing with older buildings, an initial budget based on predictable cost-per-square metre rates found in standard pricing guides is unlikely to be accurate. Repairs will often be localised and specialised in nature, and it is not until the architect or surveyor has made a careful inspection of the building that it is possible to produce the first indicative budget. The initial budget accuracy is dependent on the how closely the building has been inspected. Clients therefore who are prepared to pay for detailed inspections and **condition surveys** often benefit by having much better cost certainty and smoother contracts. The architect or surveyor will need to make as thorough inspection as is possible which may involve:

- Lifting floor boards or panels
- Lifting manhole covers
- Inspecting roof lofts and building voids
- Making trial holes (these may need listed building consent).
- Making notes and observe cracking and decaying fabric
- Using damp meters (though their value can be limited in an old building void of damp proof membranes and damp courses)
- Lifting tiles or lead flashings to establish condition of the underlay and structures beneath
- Going up ladders, scaffold towers or cherry pickers to gain access to higher walls and windows, valleys, roofs parapets etc
- Using binoculars to inspect roofs and higher walls
- Using a screwdriver or pen-knife to check decay of timber joinery
- Employing specialist trained abseiling 'wall walkers' or steeplejacks to survey or photograph high walls, church towers, chimneys etc
- Using endoscopes to probe cavities without making large holes
- Using thermal cameras to identify trapped moisture or holes
- Tapping renders and plasters to find where they have lost their key to the background fabric
- Using specialist dogs to sniff out dry rot and other moulds
- Taking detailed photographs and studying them carefully off site when there is more time

- Specialist ground penetrating radar surveys to establish voids or archaeology, say in a church floor where there may be hidden tombs. Specialist ultrasound scanning techniques to establish if old beams are hollow following beetle or dry rot attack
- Using laser cloud surveys to identify defects. These are very detailed surveys which can look like photographs but are measurable. They can also be used to create an accurate three-dimensional model of a building from which complex structural problems can be identified. The laser surveying instrument can measure into high up spaces and be put on a pole to view roofs and voids which are hard to see into
- Taking samples of mortars and renders etc. to establish mixes prior to specification
- Employing specialist accredited conservation architects to advise and carry out desk evaluations of the significance of the building using desk top methods and site investigations
- Archaeological trial holes and desk top studies can also be carried out by an archaeologist
- Employing specialist conservators who can advise on the repairs and suitable budgets required for stone, including carvings, sculpture, floors, encaustic and ceramic tiling, paintings, decorative plaster, furniture, gilding etc

Investigative works clearly need to be done safely, and the architect or surveyor undertaking this work will need to ensure s/he has safe access to undertake the work. An initial assessment may need to be undertaken and the risks evaluated and mitigated. The use of ladders will almost certainly be needed for most contracts, but these need to be properly secured and used for limited times only. Where there are safe anchor points harnesses can be used. It is also sensible never to undertake detailed investigations on your own in case of an accident.

On most projects the investigations are less intensive than would be ideal, and when suggesting an initial budget to a client it is prudent to allow a large contingency for work which is not yet visible. Initial contingency allowances can be as high as 30% of the contract value, and are not unusual when no investigations have been carried out. This can be reduced as pre-contract investigations allow accurate assessments, or increase if the building is very dilapidated and not easy to inspection.

4. Investigative contracts

Where initial investigations are extensive and require attendance of a contractor to open up the building or provide access safely, then it often becomes sensible to run an investigative contract prior to the main works. The client however may need persuading of the advantages of paying for preliminary investigative contract. Many experienced clients recognise the advantages of undertaking extensive investigations prior to signing a main contract. Many grant-awarding bodies such as Heritage Lottery Grants normally award grants in at least two stages to ensure that the risks are assessed and investigations are carried out prior to the award of funds for the main contract.

5. Ecological studies

Old buildings are often homes to more than just people, they can be the dwellings of many species. In general the older the structure, the greater chance that it has become a home to wild-life. Bats often roost in cracks and crevices in old roofs, newts winter under old floor boards, owls may build nests in redundant farm buildings, swifts nest in the eaves of the structure. The Conservation of Habitats and Species Regulations 2010 and other legislation protects many species, making it a criminal offence to disturb them. Unless extremely confident that there are no protected species present, it is good practice to get an ecologist to evaluate the likelihood of protected species being located in a building before any work starts.

It is sensible to carry out the initial scoping study as part of the initial feasibility investigations, as the design may be influenced by the need to protect any habitats in the building or mitigate the effect on wildlife of a design, for example by providing bat boxes or leaving an existing home in a loft for bats undisturbed. Following the scoping survey, often more detailed surveys are needed to confirm the findings, and these may need to be done at the appropriate time of year when the species is visible. If any protected species are present, mitigation measures will probably be needed or precautions taken not to disturb the protected species, and a licence obtained from the Department for Environment, Food and Rural Affaird (DEFRA). In these cases an ecologist will be needed to make the licence

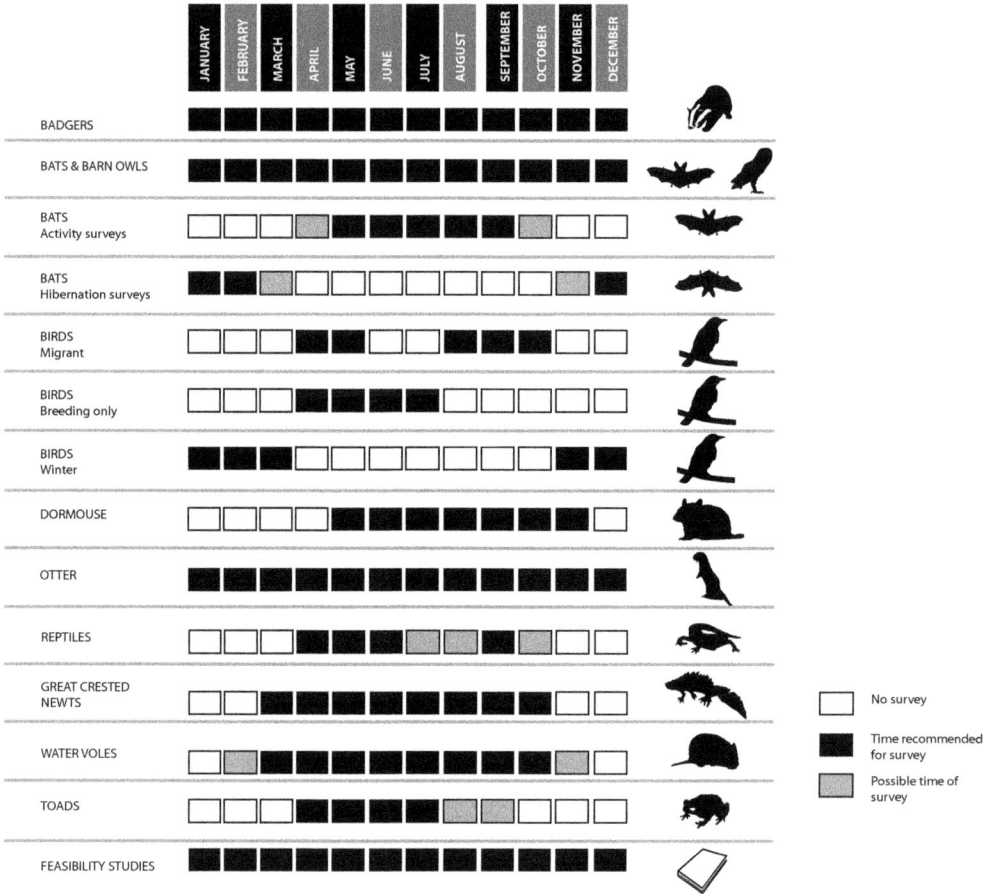

	JAN	FEB	MAR	APR	MAY	JUN	JUL	AUG	SEP	OCT	NOV	DEC
BADGERS	■	■	■	■	■	■	■	■	■	■	■	■
BATS & BARN OWLS	■	■	■	■	■	■	■	■	■	■	■	■
BATS Activity surveys	□	□	□	▨	■	■	■	■	■	▨	□	□
BATS Hibernation surveys	■	■	▨	□	□	□	□	□	□	□	▨	■
BIRDS Migrant	□	□	□	■	■	□	□	■	■	■	□	□
BIRDS Breeding only	□	□	■	■	■	■	□	□	□	□	□	□
BIRDS Winter	■	■	■	□	□	□	□	□	□	□	■	■
DORMOUSE	□	□	□	□	■	■	■	■	■	■	■	□
OTTER	■	■	■	■	■	■	■	■	■	■	■	■
REPTILES	□	□	□	■	■	■	▨	■	▨	▨	□	□
GREAT CRESTED NEWTS	□	□	■	■	■	■	■	■	□	□	□	□
WATER VOLES	□	▨	■	■	■	■	■	■	■	■	▨	□
TOADS	□	□	□	■	■	■	■	▨	▨	□	□	□
FEASIBILITY STUDIES	■	■	■	■	■	■	■	■	■	■	■	■

Legend: □ No survey; ■ Time recommended for survey; ▨ Possible time of survey

FIGURE 9.1 Protected Species Activity Calendar

application and advise on mitigation measures. Often work has to be timed to avoid when the protected species is not breeding or is dormant. For example works stripping roofs may need to be timed so that it avoids the bat breeding season in the summer.

6. Asbestos

Asbestos fibres can be fatal if breathed in, and they are found in a surprising number of building products from vinyl floor tiles, lagging of old heating pipes and many wall and ceiling boards. Asbestos is often a hidden risk in older buildings, as it was used in paints, floor and ceiling finishes and to lag pipes etc. The likelihood of asbestos in an early medieval church is probably low but it could be high in an Edwardian building when asbestos was being used more regularly. Often the asbestos is only a danger if the relevant asbestos-bearing product is broken or affected by building work.

The Control of Asbestos at Work Regulations 2012 (CAWR) require those bodies with responsibilities for the repair and maintenance of non-domestic premises to find out if there are, or may be, asbestos containing materials within them by carrying out an asbestos management survey. It also requires them to record the location and condition of such materials and then assess and manage any risk from them, including passing on information about any asbestos bearing material, its location and condition to anyone liable to disturb them, such as builders.

The building owner is legally bound to have a register of the locations of the asbestos, and prior to starting work on a project the contractor will need to see it. Where alterations and demolition are taking place, the owner will need to employ a specialist asbestos surveying company to undertake a demolition/alteration survey to establish if there is asbestos in the fabric, and if so what type it is. They will also advise about management and disposal of any asbestos which is located. The law requires that these surveys need to be carried out before any demolition or alteration work starts on site. It is good practice to examine the Asbestos Register early in the commission, as the cost of asbestos removal is high, and its presence and location may influence the design and specification.

7. Structural engineers and other consultants

It is important when working on listed buildings that the rest of the consultant team has a firm understanding of conservation principles and seeks to undertake the design of their aspects of the project in a sympathetic way. To ensure the consultant team approaches the project sympathetically, it is often good practice for the architect to write the briefs for the consultants. In these briefs, the architect should bring to the consultant's attention the significance of the fabric, where special care and attention is needed, and any risks that cannot yet be mitigated or need investigation by specialists.

Service engineers need to understand it is not always acceptable to run their services in the most efficient way or make holes without consent through historic fabric.

It is wise to employ a structural engineer with experience of working on similar historic buildings. Engineers designing new buildings can fully calculate the weight and loads imposed on the structure, but this is not always possible when working on an existing building, as one cannot always rationalise where loads are transferring or establish exact weights for existing fabric. Conservation engineers therefore often need to be able to make judgements based on experience supported by appropriate calculations.

There is now a Conservation Accreditation Register for structural Engineers (CARE) which is administered by the ICE and IStructE, and recognised by English Heritage, Cadw, Historic Scotland and NIEA.

The Chartered Institute of Building Services Engineers do not currently run a conservation accreditation scheme, but they have a heritage group for members with special interest in historic building services.

8. Quantity surveyors and cost control

The assistance of a quantity surveyor when working on an historic building can seem unnecessary when the value of the contract is relatively small, but their use can be extremely valuable to the smooth running of the project. It has to be recognised that most refurbishment and conservation projects will involve a large number of variations that need to be measured and agreed with the builder. Indeed it is not uncommon for half the specified work to be varied as the building is opened up, and scaffolding provides access for close inspections. Having a pricing document that allows for this level of change and enables easy evaluation of claims. The assistance of a quantity survey to help cost monitoring will save the architect or surveyor considerable time and possible disputes.

The use of the quantity surveyors' normal RICS approved Standard Method of Measurement is not normally successful on a conservation based project which requires a more descriptive item based specification, and which therefore ensures the contractor has a thorough understanding of the work needed.

Where bills of quantities are used, the individual clauses often need to be expanded to suit different situations and locations where similar activities are taking place. The use of a standard description and rate for a particular activity for the project may not fully describe what is needed in a particular location, and this is likely to lead to claims from the builder. For example, there may be a standard clause for measured areas of plaster repairs, but the contractor will need to know if they comprise a whole set of small repairs, or whether they form a larger area, as the small areas will be more time consuming. The backgrounds to these plaster repairs may also need different amounts of preparation depending where they are, including extensive dubbing out, or raking out of brick walls or making good to rubble walls.

On smaller conservation projects the pricing document is often a schedule of activities or a priced specification and drawings that an experienced architect or surveyor can produce themselves. The role of an experienced quantity surveyor is to weigh up the risks and allow sufficient funds for the unexpected without making the project seem ridiculously expensive to the client. The inexperienced cost advisor, be it a quantity surveyor or inexperienced architect, might only allow for the work identified in the specification following detailed site investigations, and make modest contingencies; they will ignore the consequences and costs of delays to the project caused when additional work is identified, and the time waiting for variations of the consents. This will normally lead to a difficult, time-consuming project for the professional team and an unhappy client. The over cautious cost advisor will, on the other hand, build so large a contingency into pricing documents for hidden items that it is likely to kill the project, or give licence to a sneaky contractor to make excessive profits on variations.

9. Choice of contract for conservation projects

There are many ways of running a building contract; some forms of contract are more suitable for refurbishment and conservation work than others.

Many conservation architects have built up a relationship of trust with reliable specialist contractors, and favour using negotiated contracts with these firms. They feel that this is a secure way of achieving quality on projects involving listed buildings. However many clients, including publicly funded bodies, will not normally accept this method of procurement unless a very good case can be made. They need to be seen to get best value for money, and prefer to use a method which includes competitive tendering.

Traditional single stage tendering is still the most common form of procurement accepted by most clients. Nearly always, the lowest tender is accepted, and often the cheapest contractor has made a mistake in his pricing, or spotted a way of capitalising on a weakness in the tender documentation. The contractor will often be seeking to recover costs when variations occur, and normally there is plenty of scope to do this with a refurbishment or conservation contract. The contractor will be able to exploit the inevitable changes to the scope of work if the fabric is being opened up for the first time as work proceeds. In parts of Europe it is normal to ignore the lowest price and often the second lowest price is accepted. The experienced architect or surveyor however can mitigate this by anticipating variations in their pricing documents, and a good quantity surveyor can also be used to comment in detail on the tender returns.

Two-stage tendering is sometimes used with a competitive first stage and a negotiated second stage. It can be useful on conservation projects where the architect wants the contractor to competitively bid for an initial investigative contract, provide competitive rates for his preliminaries and overheads, as well as providing rates for all the expected types of work they are likely to undertake. The knowledge gained during the initial investigative contract is used during the second negotiated stage to agree the main contract and make it fit the client's budget. This method is flexible and tends to make for a smooth running contract, and is useful where there is no possibility of exceeding the overall budget, for example with a charity or church with very limited funds and assets.

Cost-plus contracts can be useful when close control of the works is needed, or the scope cannot be fully identified. They can also be useful for long term historic repair contracts, where a contractor works for a fixed level of profit, and an agreed percentage of any materials or services s/he supplies. An example of this may be a project involving re-pointing a very large historic building where there is a fixed budget each year and the work will take years to complete.

Design and Build contracts will not normally give the contract administrator sufficient control over the materials and workmanship used on a sensitive historic building and therefore should be avoided.

For larger refurbishment contracts where there is a large amount of work requiring specialists, craftsmen management contracting can be used. In this form of contract the main contractor is paid to manage and programme packages of a series of specialist subcontractors. It may be suitable for a project repairing the façade of a large listed building needing specialist masons, cleaners for external sculpture, repairs to historic metal windows, and historic iron rainwater goods etc. The documentation needs to be high quality, and this form of contract often reduces the role of the architect and professional team.

10. The form of contract

The main reasons for choosing a particular form of building contract for conservation and refurbishment projects are no different than that for new buildings. The principle choice is driven by the need for the form of contract to be appropriate to the size and scope of the works, and will be driven by the time taken, cost of the works and the build quality.

Most architects will use the JCT suite of contracts, Minor Works, Intermediate and Standard, depending on the value of the contract, all of which are suitable for conservation and refurbishment projects. Special consideration in the choice of contract when working on existing historic buildings may also be driven by the following:

- **Phased handovers**

  When working on existing buildings there is often a need for a phased handover of completed parts of the project, however the JCT Minor Works Contract is designed for projects with only one phase, and there is no mechanism to manage a phased project, therefore an Intermediate or Standard contract should be used.

- **Specialist tradesmen**

  Historic restoration projects also regularly involve selecting specialist tradesmen who are named in the tender documents as required to work on the project. Named specialists may include stone masons, conservators etc., who have been selected to do a particular skilled job. They are often tradespeople the architect trusts, and in the intermediate and standard contracts, these choices can be imposed on the contractor.

- **Special control measures**

  When the architect needs to detail special control measures that may have an effect on the progress of the works, one needs to use the intermediate or standard contract. These control measures could include special samples approval procedures, or periods of time in the contract needed for archaeological investigations. The control of critical works is often described in detail in the specification. The Minor Works Contract is more difficult to use where one wants these controls, as unlike the Intermediate or Standard contracts there is no mechanism for dispute resolution.

The Association of Consultant Architects also publishes a well-regarded form of contract used by some conservation architects, as well as a partnering contract.

11. VAT and listed buildings

Prior to the 2012 budget there were concessions for owners of listed buildings on the amount of VAT chargeable for alterations that had listed building consent (or scheduled monument consent). This relaxation of the tax rules has been removed in the 2012 Budget. VAT now applies to all works to listed buildings in the same way as it is to unlisted buildings.

Listed churches can recover some of the VAT paid on their maintenance and repair of their fabric including clocks, electrical works, decorations, heating, organs and professional fees, as well as approved alterations by applying to the DCMS (Department for Culture Media and Sport) who run a scheme called the Listed Places of Worship Grant Scheme (LPWGS) (see www.lpwscheme.org.uk). The scheme has a fixed budget for the year, and the amount recovered is dependent on the level of claims in any quarter of the year.

5% Reduced rate

As with unlisted buildings there is a lower 5% VAT rate applicable to works on listed buildings under the following conditions:

- Conversion of buildings which result in a change in the number of dwellings, including from non-residential uses, such as agricultural barn conversions
- Renovation of dwellings which have not been lived in for at least two years
- Installation of some energy saving measures, including solar panels, ground source heat pumps, draught proofing to windows and doors

Zero rated works

Some adaptions for disabled persons are eligible for a zero VAT rating in listed buildings as with unlisted buildings.

12. Specifications for conservation projects

Many architects rely on standard National Building Specification (NBS) clauses to build their specifications, as this is a very reliable and authoritative source for modern construction, but it currently lacks many clauses based on historic and local vernacular construction techniques. Some architects take NBS clauses and adapt them, but specialist conservation architects will often develop over time their own specification clauses suitable for conservation work. The clauses may reflect local building techniques or in-depth knowledge of the construction techniques of a particular period.

When preparing a specification and details for a conservation project, it is clearly important to understand thoroughly the materials and methods used to construct the building. Historic fabric can be badly damaged by architects specifying the wrong products when they do not understand the ways historic materials behave in practice. The most common mistake is the use of modern materials like cement, modern plaster and standard plastic paints which don't allow the existing fabric to breathe and let moisture escape safely. Pre-1919 buildings were not constructed to be air or moisture tight, and if modern vapour-proof technology is brought into old buildings for missing damp proof membranes and vapour barriers, the original fabric is likely to be damaged. Trapped moisture will often lead to decay by insects, fungi and chemical reactions.

A thorough understanding of the forms of construction used in the old building is therefore necessary before the architect writes his/her specification and prepares details. There are books and technical guides available on the materials and construction methods suitable for historical work, listed fully under Further Reading, which include:

- *English Heritage Publications.* Since 1988 Ashgate Publications, on behalf of English Heritage, have produced a growing series of books with specific practical guidance about conservation techniques to be used on historic buildings in a series of Practical Conservation Guides. The topics currently covered include metals, timber, stone, mortars and renders, and glass. They give the reader a good understanding of the history, use, recording and conservation of historic materials in different situations, as well as practical advice about their repair. They are considered essential reference guides for specialists, but they would also be a useful tool for the young and emerging conservation architect.
- *SPAB's Technical Guides* are a series of practical guides on the conservation and repair of various traditional materials such as *Patching Old Floorboards*; *Rough Cast for Historic Buildings* and *Removing Paint from Old Buildings*, as well as guidance for owners of historic houses. SPAB also reprint a classic book of the 1920s by A.R. Powys *The Repair of Ancient Buildings*, which can still provide practical advice to the modern conservation architect.
- *The Journal of Architectural Conservation* is an international journal providing guidance on policy, recent practice, technical developments and academic research. The publishers Donhead have also published a series of technical books on subjects such as stone conservation and the maintenance of historic buildings.
- *The Institute of Historic Building Conservation* is a body for building conservation practitioners and historic environment specialists covering a wide range of disciplines including architects, planners, surveyors and engineers etc. They publish papers and keep a good technical biography of published sources on specific conservation issues, types of traditional buildings and materials (see www.ihbc.org.uk).
- *Cathedral Communications Ltd* publishes an annual *Building Conservation Directory* that is a useful directory for specialist conservation companies as well as including short papers on building conservation issues. They also run a useful website www.buildingconservation.com, where they re-publish the articles from their directory.

Where to find advice

The choice of materials and techniques can be a tricky decision, and specialist conservation architects learn to exploit the knowledge and experience of others to make the right choices. It is important when dealing with historic fabric to acknowledge limitations and take advice when uncertain. Experienced conservation architects are often willing to share experience with others working on historic buildings, be it informally or via various forums and CPD events, or as a specialist consultant on a particular project.

Manufacturers of traditional materials such as lime and distemper paints are normally extremely helpful, and if asked, will provide good guidance from their expertise on specific problems. Some have technical representatives who can visit a project and advise. A wise conservation architect will always recognise the skills of specialist tradespeople in guiding them to the right choices. SPAB also run a part-time technical helpline which is largely there to advise owners of historic buildings that can be useful (see http://www.spab.org.uk/advice/), while Conservation Officers and the staff of English Heritage are also a source of expertise that can be very helpful.

Specification clause writing

The individual specification and schedule of work clauses need to be written in sufficient detail to ensure that the contractor has a clear understanding of what they need to do, with what materials and where. A clear, unambiguous writing style will help avoid incorrect removals, timing or use of materials. The contractor must be made fully aware exactly what to do in a specific location and when. Normally clauses need to include information as follows:

- Location of the work
- Details of any preparation needed and the backgrounds
- The materials to be employed
- The precise methods of use and timing
- The quality of finish expected
- Any item of specific approvals or inspections needed by the architect or surveyor

It is good practice to break down each clause into its separate components and insist that each individual component is priced separately. Compiling the clauses this way will help with cost control during the contract as and when variations occur. Because the clauses are broken down into smaller parts, it is possible to deduce the specific rates for individual activities, which often vary. For example:

- Remove 10 ridge tiles
- Carefully remove the first 10 courses of peg tiles beneath the ridge tiles and set aside for re-use
- Remove battens and counter-battens underneath
- Allow to lift and replace a provisional quantity of 10m² of defective boarding with 100 × 12mm tongued and grooved softwood boarding
- Supply and fix new 25 × 38mm preservative treated softwood battens and counter-battens with 65mm stainless steel nails
- Check soundness of salvaged peg tiles with the architect
- Re-tile the roof with the salvaged tiles allowing for 15% replacement to match, using 32mm × 4mm alloy pegs

Clauses should also not leave much to the contractor's discretion, as this may well lead to disputes. Items which are not specific are to be avoided, such as 're-point all defective pointing'. This sort of clause can easily lead to disputes about the extent of defective pointing needing repair. It is far better to put in an estimated area requiring pointing and agree to the exact extent to be done on site with the contractor when the scaffolding provides access.

Keeping on top of variations through provisional quantities and hourly rates

Working on a historic building can be a process of exploration, which can make contract administration challenging. Often it is only when the contractor starts on site and coverings are lifted that it can be established how much fabric needs repair. Original and significant elements of structure may even be found hidden behind coverings, which require the design to be modified. To manage the changes needed once a project starts on site, conservation officers often keep a close interest and may visit regularly. Architects therefore need to be careful when writing a specification to cope with this level of uncertainty if they don't want to find the cost of the project rising beyond the client's expectations, as variations can be used by the contractor to increase profits.

Provisional sums could be allocated to cover the costs of unforeseen events, but these are open to being exploited by the builder to increase their profits. It is better practice to make an intelligent guess about what might be encountered, and then describe it and give estimated allowances of the repair in the form of a provisional quantity. This will establish rates for the work that may be encountered, and where a variation is needed, the work can be re-measured on site and the cost adjusted. For example the area of lath and plaster work repair on a ceiling may be estimated in pricing documents, however only once the scaffold is erected can an inspection take place, to agree the exact areas.

Claims for extensions of time can also be resisted if one allows sufficient provisional quantities in the documentation, and they are identified where they are likely to be needed. The contractor should have allowed for the work described by the provisional quantity within their tender. The variation therefore should have no effect on his/her programme. This can be important when dealing with slow-drying repairs, such as lime renders. If one only describes the work that is obvious, and allows a provisional sum or a contingency for extra repairs, then if there is a lot of slow-drying lime render work added by a variation, the contractor could have a legitimate claim for an extension of time, and the costs of additional preliminaries whilst the lime dries. However if the 'contingency' had been hidden in the form of a provisional quantity, then the contractor should have allowed for the extra time to do the work in his original programme.

Where variations occur for items that are hard to quantify, valuing them can be hard to judge, such as for example repairs to joinery. It is therefore sensible to require tendering contractors to competitively provide hourly rates for the trades needed on the project. It is also prudent at the tendering stage to require the contractor to make a clear statement of the additional mark-up required to cover overheads and profit on any materials needed in the event of a variation.

Contract period

The timing of individual items of work can be a major consideration when using traditional materials such as lime putty mortars or lime renders. The lime will need time to carbonise and set, and this will be affected by the temperature and climatic conditions. The contractor cannot be expected to carry out repairs involving lime putties in the coldest winter months, as the lime is unlikely to go off. It is therefore important to bring the contractor's attention in the specification to the likelihood of delays whilst traditional materials dry or set, in order to ward off extension of time claims.

The climate sensitive nature of many traditional materials may also determine the appropriate time in the year to undertake the building contract. For example, a project involving external lime rendering will need to be carried out when the weather is unlikely to be frosty, and therefore the contract administrator should carefully consider the likely programme of building works when determining the starting date for the contract.

Rather than relying on the contractor to state the length of contract required to do the work, in conservation based projects it is normally important for the architect to state the length of the contract. Contractors who are allowed to state the length of time they want for the contract will often underestimate the time needed, in order to keep their preliminary costs competitive, and might look for reasons why they have been delayed to increase the length of the project. This can lead to extension of time claims or loss of quality from rushed work.

The contract administrator should require the contractor to produce a contract programme before accepting the tender. The programme should be studied to identify that climate sensitive activities are carried out in the appropriate months of the year, and that the contractor has allowed sufficient time for sensitive activities.

Time allowances for investigations and archaeology

When working on a listed building there are often conditions applied to the listed building consent for an archaeologist to be in attendance when parts of the construction are opened up or holes dug. This is called an 'archaeological watching brief'. The client will need to pay for the archaeologist to be on hand while the building is opened up and whilst construction work is stopped for the recovery of archaeological artefacts, and records made of any finds. For example, in carrying out drainage works in or around a listed building, the archaeologist may be standing over the digger watching any material being uncovered. The work may be stopped at any time whilst the trench is inspected. The delay to the project can lead to an extension of time claim and resultant increase in costs. To avoid claims, allow an adequate provisional sum in the specification for lost time whilst the archaeologist is recording and the contractor is unable to proceed.

By talking to the archaeologists prior to writing the specification it is possible to get an indication of a sensible time allowance to be incorporated into the contract documents for delays caused by archaeological investigations.

It may even be prudent to commission a detailed desktop study prior to the building contract to establish the likelihood of encountering archaeological finds. This can be done by an experienced conservation architect or an archaeologist. An initial analysis can normally be done by studying the Local Authorities' Heritage Environmental Records (HER's) in the vicinity of the property to identify any archaeological finds that have been recorded locally. Historical and cartographical sources are also used to build an understanding of the development of the site and history.

Specialist artists and tradesmen

It is often necessary to commission specialist tradespeople for elements of the project. These could be conservators, specialist masons etc, and sometimes they are directly employed by the client. Their work would be controlled in the contract by the clauses dealing with 'artists and tradesmen', and like the archaeologists, if they are delayed or fail to finish on time, they can cause delays to the project completion date, resulting in claims from the contractor. Having a thorough understanding of their programme and the risks for delay is therefore important, and expressing this clearly in the contract documents can help contractors allow sufficient time in their contractual programme for their attendance, and thus avoid claims.

Using samples to achieve the desired result

New work or repairs to an old building often needs to match the existing fabric in colour or texture. Specifying exactly the right materials and techniques to achieve a desired result is very difficult. One way to overcome these problems is to require the contractor to build sample panels close to the fabric to be matched. Several samples may be needed to provide a range of examples, and to allow the best to be selected. Often the architect may need to be on site when the craftsman is making the sample so they can agree the mixes and textures. It is therefore good practice to include in the specification sufficient sample panels as are needed for critical elements of the fabric, and specify how much notice is needed to inspect them.

Often the Conservation Officer or the planners will need to agree to the sample, and this can delay the project further. It is therefore advisable to state in tender documents the length of notice the council officers require to visit and process their agreement to the sample.

13. Contractor selection

The choice of the right contractor is always important to a successful project. The skills needed when working on historic buildings can be very different from the contracting skills needed to deliver a modern high-tech building. A bricklayer who is used to using ready made quick drying cement mortars, and prides himself on speed, may not be the right tradesman to deliver a careful repair to an old lime mortar wall. The vast majority of repairs undertaken each year on pre-1919 buildings are carried out by general builders who often use inappropriate materials and techniques, so finding contractors who understand how to repair historic fabric is important.

There are now many specialist contractors who can show a track-record of historic restoration. Visiting finished projects can establish if the builder is skilled in the areas needed on your project. It is worth following up references from other clients, and especially other architects and surveyors.

Contractors who specialise in historic repairs often still keep skilled craftsmen directly employed rather than relying on teams of subcontractors. Some others have close reliance on specialist subcontractors who regularly work for them.

Checking that the craftsmen employed are qualified is also useful. Since 1995 the Construction Skills Certificates (CSCS) has been established with some 1.5million card holders in the construction industry. There is now the CSCS Heritage Skills Card which the National Heritage Training Group and English Heritage have encouraged and require for any tradesmen working on their properties. A list of the competencies for various skills cards can be found on the www.cscs.uk.com website, including bricklaying, carpentry, craft masonry, earth walling, painting and decorating, plastering and fibrous plastering, roof slating and tiling, stone carving, tile and floor laying, and lead work. There are also cards for site managers, agents, and Conservation Controllers. There are also Construction Related Occupations (CRO) cards for skills where there is no nationally recognised training system. Trades covered by this card include architectural ironwork and dry stone walling. Unfortunately, the current take up of both of these schemes is still very limited and unlikely to grow unless architects and clients insist that tradesmen working on their projects have cards to demonstrate their skills.

14. The importance of insurance

Historic buildings, because of their design and construction, are intrinsically more susceptible to accidental damage. They are especially vulnerable during building works, and many irreplaceable historic buildings have been destroyed by careless construction activities. It is very important therefore to take care through the construction period to prevent damage, particularly by flood, fire or theft. The risks for the insurance company are normally higher than for new build contracts, as the cost of repairing or rebuilding a historic building can be very high due to the specialist skills and materials needed and the unpredictable nature of surviving fabric. Insurance claims can be so expensive that it is not unusual for them to exceed the commercial value of the property. Clients often do not realise the implications of building work on their insurance, but as building professionals working on historic buildings, the architect has a duty to educate them into the risks and ensure they are adequately insured.

Many conventional insurers are nervous about insuring listed buildings during building works because they are so vulnerable to damage, but there are a number of specialist brokers and insurers such as Ecclesiastic Insurance Group who were originally set up to insure ancient churches.

It is normal when working on existing buildings for the insurance of the building to be provided by the client and not the contractor (and not in joint names). Valuing a historic building for insurance purposes can be difficult, and a contractor should not be expected to do this, whereas the client should know the value of his property and its contents.

It is also important to realise the terms and conditions of the insurance, as many household insurances policies can differ significantly, and it is wise to remind clients to check their policies and get them before work starts.

When clients are considering their insurance cover during building works they should consider whether they need:

- Cover for special high quality fixed architectural features such as special carved woodwork, Georgian door casings, Tudor staircases and period fireplaces. These all have value in the second hand architectural market and considerations should be given to insuring these agreed items.

- Contents insurance normally covers items not permanently fixed to the building including furnishings, works of art, collections and personal possessions. During building works valuable or vulnerable unfixed items are normally required to be removed. If they can't be removed, one may need to ask the client to record and provide specialist valuations for the insurers to ensure they are covered by insurance.

- Projects on commercial buildings may need cover for consequential loss and 'business interruption' or 'loss of profit' insurance. This cover provides insurance for loss of revenue from rent, visitors or loss of profit for a business which has to close.

- Cover for the cost of alternative accommodation may also be needed, as rebuilding times for historic buildings are generally longer than for modern buildings.

- Cover for any public or third party liabilities for claims of damage to persons or property, made by employees, visitors and neighbours. Normally these third parties would be excluded from the site of the works and the contractor would be required in the contract to have his own third party cover.

- Most policies provide engineering insurance to cover damage to the mechanical plant in a building such as lifts, boilers etc, but this is not always needed during a building contract.

- Historic buildings often have historically significant gardens which may have valuable garden ornaments or special landscape features, and the client may need to ensure they have appropriate insurance cover during the building works. Ornaments may be especially vulnerable during construction works from theft or accidental damage if they are close to access ways.

- The risks of terrorism may be low for most projects, but since 1993, insurance against terrorism is often excluded or limited for most commercial premises, as well as for many listed residential properties and is expensive to reinstate.

Additional premiums are often required and these can be hefty. The terms of cover are also often modified by endorsements which limit the liability of the insurer and add risk for the client. It is important that the architect communicate to the client and their insurer the risks the building may face through the construction period and the scope of the works.

Insurers regularly require special precautions to be undertaken by owners and the builders. It is therefore important ahead of preparing the specification to understand any requirements of the building insurer which may include:

- The prohibition of processes with naked flames or those likely to produce sparks
- The use of hot working permits to control where and when any operations which may lead to fires are carried out. For example preventing the use of any hot working in the last two hours of the working day so any smouldering fabric can be spotted whilst contractors are on site. Permits need to be issued by a competent 'authorised' person following a risk assessment of the process (such as the architect or fire officer). The permit should be for an agreed length of time and the risks should be mitigated. It is normal to have a written check-list to be completed by the operatives just before they do the work to ensure all mitigation measures are in place, such as having two fire extinguishers immediately available

- Strict controls on access and escape arrangements and the strategic placing of fire-fighting equipment
- The requirement for a nominated person to inspect the site at night to check for any risks
- Controls on the use and storage of solvents and flammable materials on site
- The use of alarms on scaffolds to prevent unauthorised access at night and to prevent the theft of valuable fabric like lead roofs or historic fireplaces
- The boarding of the first 3m of scaffolding to prevent criminals climbing onto or into the building.

In the event of a major accident the records of the existing building will be tested, and this is yet another reason why when working on historic fabric it is essential to properly record the existing fabric with measured surveys and photographic records before the work starts.

15. Health and safety issues

Asbestos is not the only health and safety risk associated with working on historic buildings. Care and assessments need to be taken before work starts to try and identify any risks. These may include:

- Anthrax spores in old horse hair reinforced plaster which may require workmen to work in well ventilated spaces with masks etc.
- Pigeon and birds droppings in roofs and abandoned buildings which may carry microbiological and fungal diseases. These droppings are now considered a contaminated waste and will need cleaning by professionals unless the quantities are very small.
- Old drains and cavities in old buildings may be homes to rats and other vermin which may carry diseases like Weil's disease requiring contractors to use masks and gloves and to be able to adequately wash after any contact.
- Old buildings may have rotted to a point when they are unstable, requiring precautions to be taken before work continues to shore up or prop. Sometimes the choice can be to carry out a complicated repair requiring considerable temporary structures rather than to demolish and rebuild, as it will often preserve more original fabric.
- Old hazardous materials such as lead in paintwork, arsenic in wallpaper requiring masks and ventilation before old finishes are sanded and repaired. White lead was commonly used in paintwork up until the 1970s and was only banned in 1988, and arsenic can often be in early wall papers where it was used to inhibit mould growth. Sanding off old paint or paper therefore risks spreading lead dust. Burning it off will lead to poisonous fumes that can be inhaled, and poultices and washing risk contaminating ground water. Therefore it is often better not to attempt to strip old paint finishes, but rather cover it in new paint thereby encapsulating it. This also has the advantage of preserving the evidence of earlier colour schemes and decorative schemes that can be unearthed by technical analysis.

16. Site visits

When administering a building contract it can be common to visit the site only once or twice a month to check progress and the workmanship and materials being used, however when working on historic buildings much more frequent visits are nearly always needed. Site visits are vital when elements of the structure are opened up and coverings removed to evaluate if the specification is correct, and what repairs are needed. They may require modifications to the specification and design, for example floorboards may be raised and dry rot revealed needing immediate treatment. The correct use of specified materials needs to be checked, for example to ensure the contractor is not using cement in a mortar when only lime is specified. Where new work needs to match existing fabric, samples may need to be approved requiring further visits. The skills, workmanship and performance of the contractor may need to be evaluated and this can only be done on site.

In order to advise clients about future maintenance issues it can also be very useful to visit the site and examine the condition of fabric close up from scaffolding, towers etc.

17. Recording

It is good practice when working on a conservation contract to leave records on the completion of the works of interventions and alterations made to the fabric so future generations can interpret the building correctly and make informed decisions about future repairs and maintenance. The level of detail of any recording should be proportionate to the significance of the fabric. It could include copies of record photographs and drawings before and after the works, as well as the specification of the works.

It is good practice to leave a file containing these details that could be included in the Health and Safety file for non-owner-occupied domestic buildings, or in the case of church buildings lodged in the parish archives. Owners of very significant historic buildings such as stately homes or castles may well maintain an archive where the information can be kept. The local planning authority can advise where nationally important information should be kept, and they can add this information to the national database system of Heritage Environmental Records. They may also suggest that important records are stored in county or borough archives or in the National Archives. For smaller domestic properties it is sensible to keep a copy to give to the client, explaining its significance for future owners in the hope that it is passed on.

Further reading

1. Scope of the works and conservation philosophy to be adopted

> Earl, J. (2003), *Building Conservation Philosophy*, Shaftesbury, Donhead.
> English Heritage (2013), *Conservation Basics*, Practical Building Conservation, Farnham, Ashgate.
> Insall, D. (2008), *Living Buildings; Architectural Conservation, Philosophy, Principles and Practice*, Musgrave, The Image Publishing Group.
> Seeley, N. (2006), 'Commissioning Conservation Work', in *The National Trust Manual of Housekeeping – The Care of Collections in Historic Houses Open to the Public*, Oxford, Butterworth-Heinemann.

2. Getting the fee right

> Roland, P. (2012), *RIBA Good Practice Guide: Fee Management, 2nd Edition*, London, RIBA Publishing.
> RIBA (2010), *A Guide to Working with an Architect: Repair and Alteration of Historic Buildings*, RIBA Publications
> RIBA (2010), *A Guide to Working with an Architect: Repair and Alteration of Places of Worship*, RIBA Publications

3. Budgets and initial assessments

> English Heritage (2013), *Conservation Basics: Practical Building Conservation*, Farnham, Ashgate Publishing Ltd.
> Fielden, B. (1994), *Conservation of Historic Buildings*, London, Architectural Press.
> Hoxley, M. (2009), *RIBA Good Practice Guide – Building Condition Surveys*, London, RIBA Publishing.

4. Investigative contracts

> English Heritage (2013), *Conservation Basics: Practical Building Conservation*, Farnham, Ashgate Publishing Ltd.

5. Ecology

> More advice can be found on Natural England's website www.naturalengland.org.uk
> English Heritage, National Trust, Natural England (2009), *Bats in Traditional Buildings*, Swindon, English Heritage, also available from www.english-heritage.org.uk
> English Heritage (2013), *Conservation Basics: Practical Building Conservation*, Farnham, Ashgate Publishing Ltd.

The legislation protecting species and habitats is spread across a number of acts and measures but the principle three are as follows:

> HM Government, *Wildlife & Countryside Act 1981* (as amended), available from www.legislation.gov.uk.
> HM Government, *Conservation of Habitat & Species Regulations 2010*, available from www.legislation.gov.uk.
> HM Government, Protection of Badgers Act 1992, available from www.legislation.gov.uk.
> Murphy, B., Gunnell K. and Williams, C. (2013), *Designing for Biodiversity: A Technical Guide for New and Existing Buildings* (2nd Edition), RIBA Publications
> HM Goverment, *The Conservation of Habitats and Species Regulations 2010*, available from www.legislation.gov.uk

6. Asbestos

> HM Government (2012), *The Control of Asbestos at Work Regulations 2012 (CAWR),* available from www.gov.uk.
> More advice can be found on the Health and Safety Executive website www.hse.gov.uk/asbestos.
> HM Government, the Control of Asbestos at Work Regulations 2012 (often known as CAWR), available from www.legislation.gov.uk

7. Structural engineers and other consultants

> The Register of Structural Engineers who are members of CARE can be found at www.careregister.org.uk.
> The CIBSE Heritage Group have a web site available at www.heatvac-heritage.org.

8. Quantity surveyors and cost control

> Hughes, N. (1996), *Tenders for Conservation Work: Building Conservation Directory 1966,* Tisbury, Cathedral Communications, available from www.buildingconservation.com.
> Stenning, A. and Evans, G. (2007), 'Costing and Contracts for Historic Buildings', in *Understanding Historic Building Conservation,* Michael Forsyth (ed.), Oxford, Blackwell Publishing.

9. Choice of building contract for a conservation project

> Stenning, A. and Evans, G. (2007), 'Costing and Contracts for Historic Building', in *Understanding Historic Building Conservation,* Michael Forsyth (ed.), Oxford, Blackwell Publishing.

10. The form of contract

> Luder, O. (2006), *Keeping out of Trouble: RIBA Good Practice Guide,* London, RIBA Publishing.
> Lupton S., *JCT Guides to specific contracts,* London, RIBA Publishing.
> Stenning, A. and Evans, G. (2007), 'Costing and Contracts for Historic Buildings', in *Understanding Historic Building Conservation,* Michael Forsyth (ed.), London, Blackwell Publishing.
> RICS(1998), *SSM7 Standard Method of Measurement of Building Works,* London RICS.

11. VAT and listed buildings

> Potts, J. (2013), *Vat Update: Conservation Directory 2013,* Tisbury, Cathedral Communications, available from www.buildingconservation.com.
> Further information on VAT can be found on the HM Revenue and Customs website at www.hmrc.gov.uk.
> Further information, and information on other tax matters relating to heritage assets can also be found on the English Heritage website at www.english-heritage.org.uk.

12. Form of specification

> English Heritage (2013), *Conservation Basics: Practical Building Conservation*, Farnham, Ashgate Publishing Ltd.
> Potts, J. (2013), *Conservation and the Role of the Archaeologist: Conservation Directory 1994*, Tisbury, Cathedral Communications, available from www.buildingconservation.com.
> Lithgow, K. (2006), 'Building Work: planning and Protection', in *The National Trust Manual of Housekeeping; The Care of Collections in Historic Houses Open to the Public*, Oxford, Butterworth-Heinemann.
> Seeley, N. (2006), 'Commissioning Conservation Work', in *The National Trust Manual of Housekeeping; The Care of Collections in Historic Houses Open to the Public*, Oxford, Butterworth-Heinemann.

14. The importance of insurance

> Brown, D. (2007), *Insurance for Contracts: Conservation Directory 2007*, Tisbury, Cathedral Communications, available from www.buildingconservation.com.
> English Heritage (1994), *Insuring Your Historic Buildings – House and Commercial Buildings Parts 1 & 2*, London: English Heritage.
> Further advice can be sought from The Historic Houses Association, whose website is http://www.hha.org.uk and the Listed Property Owners Club, whose website is www.listedpropertyownersclub.com.
> Individual insurers provide guidance for hot working permits. An example of a hot working permit can be found on the Ecclesiastical Insurance Groups website at www.ecclesiastical.com.

15. Health and Safety issues

> Jamieson, N. (2009), *RIBA Good Practice Guide: Inspecting the works*, London, RIBA Publishing.
> More information is also available of common hazards from the Health and Safety Executive website on www.hse.gov.uk.

GLOSSARY

Architects engaged to work on historic buildings need to understand the standard technical terms used by other stakeholders involved in the maintenance and management of heritage assets. Various publications provide glossaries, but the *British Standard 7913:2013 Guide to the Conservation of Heritage Assets* should be seen as providing the authoritative guide to industry standard terminology.

Alteration	Work carried out to change the function of a building, or which modifies its appearance.
Authenticity	Authenticated characteristics of a building, revealing its historical and cultural origins.
Breathable	A term used to describe any building material or form of construction that allows water vapour to permeate through it, and will therefore absorb or release moisture with changes in the relative humidity in the surrounding environment.
Condition Survey	Inspection to assess condition of a building.
Conservation	The process of managing change or repairing a historic fabric in a way which sustains its significance and minimises interventions (see Chapter 3).
Conservation Area	An area of special architectural historic interest whose character or appearance a local planning authority wishes to preserve or enhance. The area will have identified boundaries and be designated under the Planning (Listed Buildings and Conservation Areas) Act 1990 (see Chapter 5).
Conservation Management Plan (also sometimes known as a Conservation Plan or Heritage Plan)	A document produced to define what is significant about a historic building or place, thus establishing principles for its maintenance and management to maintain its significance. It can also be used to inform decisions about proposed changes.
Conversion	Change of use of a building.
Consolidation	The stabilisation or repair of historic fabric with minimal replacement.
Competent Person	A person who has sufficient expertise to give advice, alter or repair historic fabric without destroying its significance without good reason.

Cultural Significance	The value people associate with an object, building or place. It is a reflection and expression of their knowledge, beliefs, and traditions, or their understanding of the beliefs and traditions of others.
Curtilage	There is no strict technical definition of Curtilage but it is the area around a listed building which may include other structures associated with, and ancillary to the listed building (see Chapter 5)
Evidence Value	The capacity of a building or place to yield evidence about its past.
Ecclesiastical Exemption	The provision under the Ecclesiastical Exemption (Listed Buildings and Conservation Areas) (England) Order 2010 which exempts 5 large religious denominations from having to get listed building consent for their listed places of worship, and allows them to administer their own alternative systems of consent.
Fabric	The materials used to make and furnish a building or place.
Faculty	The formal permission issued by Diocesan Chancellors of the Church of England to allow works to be carried out to places of worship, including listed buildings in accordance with the provisions of the Ecclesiastical Exemption (Listed Buildings and Conservation Areas) (England) Order 2010.
Harm	Inappropriate interventions which diminish the historic, architectural, or cultural significance of a building fabric or place.
Heritage Asset	An object, building or place which has significance.
Heritage Assessment or Statement	A heritage assessment is a document to convey information about a building's history and significance, as well as where any impacts on the significance of proposed changes is recorded. It is normally included in the design and access statement or as a separate document. 'heritage assessment' is also sometimes referred to as a 'historic statement' or 'heritage impact assessment', or 'heritage assessment', and when dealing with a church as a 'statement of significance' (see also impact assessment).
Historic Environment	An environment resulting from the interaction between people and places through time, including surviving structures, buried archaeology, or evidence of deliberately planted or managed flora.

Historic Environment Record	A public, map-based data set, primarily intended to inform the management of the historic environment. The Heritage Gateway (www.heritagegateway.org.uk) is a significant access point for around 60% of the English Historic Environment Records held by public authorities.
ICOMOS	International Council on Monuments and Sites is a non-governmental international organisation dedicated to the conservation of the worlds monuments and sites.
Impact	Conservation professionals often use this word to describe the effect of change to a significant built fabric.
Impact Assessment	A written assessment of the impacts of proposed changes on a heritage asset. Normally a section within a heritage assessment.
Integrity	The wholeness and honesty of a historic object, building or place.
Intervention	Any act which physically alters the fabric of an object building or place.
Listed Building	A building, object or structure which is of national significance and has been included on the statutory *List of Buildings of Special Architectural and Historic Interest* (as prescribed in the Planning (listed Buildings and Conservation Areas) Act 1990. See Chapter 5.
Listed Building Consent Application	An application to a local planning authority for works to demolish or alter, extend a listed building in any manner which may impact on the character of the buildings special architectural or historic interest.
Local List	There are two types of 'local list' which planners can refer to as follows: • *'Local List of Historic Buildings'* – a list of buildings prepared by a council which have no statutory protection, but have significance, and whose loss would impact the character of the area. • *'Local List of Requirements'* – The list of additional requirements that a local planning authority can add to the National List of Requirements which are needed to make a valid planning or listed building consent application.
Maintenance	Routine work to remedy defects caused by natural decay or damage which can be carried out without changing the character of the historic building. It does not normally require the renewal of fabric, nor any

	adaptations or restoration. Maintenance of listed buildings will probably not need consent.
Natural Change	An alteration to an object, building or place without any human intervention, such as weathering. It may lead to the need for periodic renewal in order to sustain the significance of a place.
Natural Heritage	Historic habitats, ecosystems, geology and landforms which people attach value too.
Patina	The character derived from the ageing process, be it natural or man-made.
Permitted Development	Alterations and extensions which do not require a consent to execute.
Renewal	The replacement of decayed fabric with sound new material of the same type.
Repair	Work to remedy defects natural decay or damage which is more extensive than normal maintenance. It may involve renewal of fabric or minor adaptations, all to achieve a sustainable outcome. Repairs to a listed building will probably need consent.
Restoration	To return an object building or place to a known historic state. Restoration work to a listed building will need consent, which will only be given where there is a strong case for it based on evidence without conjecture.
Reversible	Capable of being reversed to a previous condition. Many approved alterations to listed buildings are designed so their impact on significant fabric is so minimal that they could in theory be reversed with minimal restoration.
Setting	The surroundings of an object, building or place from which it is seen and experienced.
Significance	The sum of the cultural and natural heritage values of a place often set out in a statement of significance or heritage statement. The NPPF definition is: 'the value of the heritage asset to this and future generations because of its heritage interest. That interest may be archaeological, architectural, artistic, or historical. Significance derives not only from a heritage asset's physical presence, but also its setting'.
Supplementary Planning Guidance	Guidance which supplies supplemetary information in respect of policies in a current or emerging plan or national planning policy
Value	The importance attached to the cultural, historic and artistic qualities of places.
Value-based judgement	An assessment that reflects the values of the person or group making the assessment.

INDEX